A U R O R A

DOVER MODERN MATH ORIGINALS

Dover Publications is pleased to announce the publication of its first volumes in our new Aurora Series of original books in mathematics. In this series, we plan to make available exciting new and original works in the same kind of well-produced and affordable editions for which Dover has always been known.

Aurora titles currently available are:

Numbers: Histories, Mysteries, Theories by Albrecht Beutelspacher. (978-0-486-80348-7)

The Theory and Practice of Conformal Geometry by Steven G. Krantz. (978-0-486-79344-3)

Category Theory in Context by Emily Riehl. (978-0-486-80903-8)

Optimization in Function Spaces by Amol Sasane. (978-0-486-78945-3)

An Introductory Course on Differentiable Manifolds by Siavash Shahshahani. (978-0-486-80706-5)

Elementary Point-Set Topology: A Transition to Advanced Mathematics by André L. Yandl and Adam Bowers. (978-0-486-80349-4)

Additional volumes will be announced periodically.

The Dover Aurora Advisory Board:

John B. Little
College of the Holy Cross
Worcester, Massachusetts

Daniel S. Silver
University of South Alabama
Mobile, Alabama

An Interactive Introduction to
KNOT THEORY

Inga Johnson
Willamette University

Allison Henrich
Seattle University

DOVER PUBLICATIONS, INC.
Mineola, New York

Bibliographical Note

An Interactive Introduction to Knot Theory is a new work, first published by Dover Publications, Inc., in 2017.

International Standard Book Number
ISBN-13: 978-0-486-80463-7
ISBN-10: 0-486-80463-1

Manufactured in the United States by LSC Communications
80463101   2022
www.doverpublications.com

Contents

A Note to the Reader

This book does not follow the design of a traditional math textbook. There are very few complete proofs included, and the exercises are not listed at the end of each section. Instead, this text is an invitation to ponder, question, create, and *figure out on your own* some beautiful mathematical results in the field of knot theory. Exercises are sprinkled in between statements of definitions, descriptions, and propositions. Some exercises are designed to introduce you to new ideas and to point you in the direction of why or how an idea is important. Other exercises guide you through the technical and subtle arguments that provide the foundation for our understanding of knots and links. Answers are rarely provided in full, but the path toward a solution or proof is illuminated.

As indicated by the word *interactive* in the title, this book is meant to be read with paper and pencil (colored pencils, string, and pipe cleaners, too!) in hand so that you can jot down your ideas, explore how a new definition applies to your favorite examples, find the answer to a question, or prove a theorem. As you are reading the text, we hope you will be in a space where conversations and collaborations with others are readily available.

A Note to the Instructor

We designed this book for a student-centered classroom where students regularly spend class time working collaboratively on problems, presenting solutions, and vetting arguments made by their peers. We have used the textbook both in a 10-week quarter and in a 15-week semester. Our classes typically have 10-20 undergraduate students, though this book could be used by a larger class of up to 30 students or even by a single group of 2-3 undergraduate research students. Indeed, we have successfully used parts of the book to introduce summer research students to the fundamental ideas in knot theory. Prerequisites for this book include a course on proof writing (which includes exposure to ideas like sets and modular arithmetic) and proof techniques (e.g., proof by induction) as well as a course on linear or matrix algebra.

Chapter 1

Playing & Building Intuition

Each section in Chapter 1 is a hands-on introduction to knots, links, and equivalence. In this chapter, we will learn about several foundational concepts in the study of knots. The *formal* definition of a knot is postponed to Chapter 2 to allow time to first play and build intuition within the mathematically rich and beautiful field of knot theory.

Informally, a **knot** is a closed loop in space. The term **closed loop** means that the loop has no loose ends, and no beginning or ending points. You can think of a knot as a knotted-up circle made of string or wire. A **link** is a collection of closed loops in space and the number of loops is called the number of **components** of the link. A link can have one component. Thus, knots are just special types of links having only one component. Note that, when we use the term 'link' in this book, we are generally referring to both knots and links with more than one component.

The examples of knots and links seen in Table 1.1 are flat drawings of 3-dimensional loops in space. A 2-dimensional drawing of a link is called a **diagram** of the link. In a diagram, the term **crossing** is used to describe a location where one portion of the link passes over another portion of the link. Crossings are identified by a short break in the drawing of the curve, which indicates that this portion of the curve is passing under the unbroken portion of curve.

Two links are called **equivalent** if they have the same number of components and they can be physically manipulated in space (rotated, bent, twisted, stretched, etc.), without cutting, so that the first link is transformed identically into the second. We imagine that our strings are highly elastic so they can be scaled up or down in size, stretched and contracted.

While playing and building your intuition with the activities in this chapter, you may come up with your own questions or conjectures. We encourage you to collect and write down your ideas and add them to the list of questions in Section 1.8. Perhaps you will create your own new open research question about knots or perhaps you will stumble upon the same questions that the founders of the field of knot theory have puzzled over for years.

One last note before you begin. In mathematics, formal proofs of theorems rely on concepts and constructions being *formally defined*. Since the formal definition of a knot is not given in Chapter 1, we will not be asking the reader for formal proofs in this chapter. Instead, we ask the reader to give an *argument* that a statement is mathematically valid. We specifically use the term 'argument' rather than 'proof' to allow for what we recognize is a tension between the desire for a formal proof with an initial lack of the formal definitions that are needed to make such a proof rigorous. An argument, in this setting, may be viewed as being less formal than a proof, but it should be as clear and complete an explanation as possible. In Chapter 2, you will notice a shift from play to formalism. Formal proofs related to the definition of a knot or link will serve as the foundation both for our play and for our proofs in the remainder of the book.

Table 1.1: Examples of Link Diagrams.

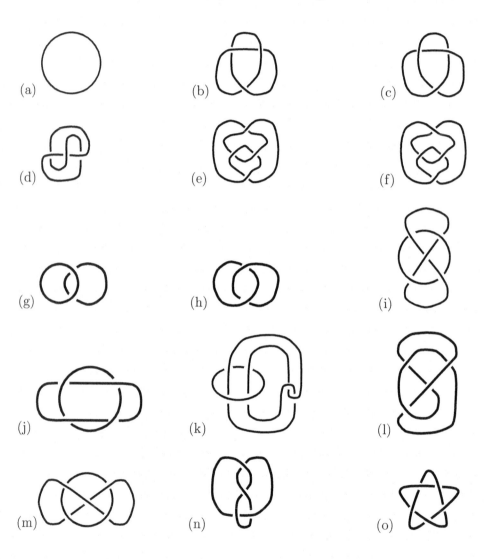

1.1 Projections, Diagrams & Equivalence

Exercise 1.1.1. (a) Identify the *knots* in Table 1.1. Build them with pipe cleaners, and determine any equivalences between different pictures. Record your findings and conjectures.
(b) Identify the *links with more than one component* in Table 1.1. Build them with pipe cleaners, and determine any equivalences between different pictures. Record your findings and conjectures.

Given a link L in space and a light source some distance away, the shadow of the link made on a plane across from the light source is called a **projection** or **shadow** of the link. Projections can look similar to the pictures in Table 1.1, but they are missing information about which is the under-strand and which is the over-strand. The curve intersections in a projection are called **precrossings**. A precrossing is said to have been **resolved** once we have selected the crossing information (that is, we have specified which strand passes over and which passes under at the crossing). Once all crossing information is determined in a link projection, the image is then a **link diagram**.

Figure 1.1.1: A knot projection in a plane, the knot in space, and a light source.

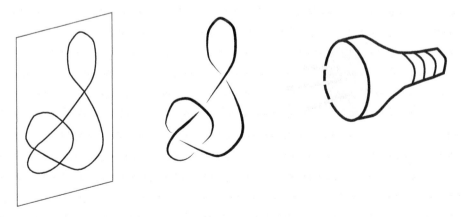

The projection of a link L onto two distinct planes in $\mathbb{R}^3$ can result in strikingly different images. Some projections are nonstandard and cannot be used to recreate the link in space, even if crossing information is included. While such projections may be useful in certain situations, they are not the projections that knot theorists typically study. (A list of specific projections characteristics to avoid will be studied in Chapter 2.) Two different projections of a link, L, can also have vastly different numbers of crossings. The following two exercises investigate the relationship between a link, its numerous projections, and the diagrams stemming from those projections.

Exercise 1.1.2. (a) Make a knot out of rigid material such as wire. Draw a projection of that knot from two perspectives that result in significantly different projections. (b) For that same knot, draw a nonstandard diagram of your knot from which the knot cannot be reconstructed. (For instance, maybe there are places in your diagram where several strands are tangent to each other or where three or more strands of the knot intersect at a single point.)

Exercise 1.1.3. Create a single rigid knot K in space such that one projection of K results in a diagram with four crossings and another projection results in a diagram with zero crossings.

Exercise 1.1.4. The knot labeled (a) in Table 1.1 is called the **unknot** or the **trivial knot**. Give an argument explaining why any knot diagram with exactly one or exactly two crossings must be equivalent to the unknot. (Hint: Draw the crossing(s) first, and then make a knot by connecting the ends in all possible ways so that no more crossings are created. Approach this task systematically so that your argument shows, without a doubt, that *all* possible ways have been considered.)

We've just seen that there are no nontrivial knots that can be drawn with just one or two crossings. In fact, the smallest nontrivial knot is a knot that can be drawn with three crossings. Any knot that can be drawn with three crossings and no fewer is typically referred to as a **trefoil** knot.

1.2   Crossing and Unknotting Numbers

Some of the links in Table 1.1 can be manipulated in space and then redrawn with fewer crossings. The **crossing number** of a link L is the minimum number of crossings needed in a diagram of L. For instance, if the crossing number of L is five, then it is impossible to draw a diagram of L that has four or fewer crossings.

Exercise 1.2.1. For each link in Table 1.1, make a conjecture about the crossing number of the link. Can you provide an argument in support of any of your conjectures? Which conjectures do you not yet have enough tools to prove?

Exercise 1.2.2. Determine the crossing number of the knot in Figure 1.2.1.

Figure 1.2.1: A complicated knot?

Exercise 1.2.3. Consider the knot projection in Figure 1.2.2. Start with an arbitrary precrossing of P and resolve it into a crossing. Imagine grasping the over-strand and pulling this strand up out of the plane of the paper. Using this visualization as inspiration, show for your projection that the remaining precrossings can be resolved so that the resulting knot is the unknot.

Figure 1.2.2: A knot projection.

Exercise 1.2.4. Use the example in Exercise 1.2.3 as a guide to write an argument that given *any* knot projection, the precrossings can be resolved to produce a diagram of the unknot. (Hint: There are two parts to this

problem. First you must *devise an algorithm* to create the desired unknot diagram. Then you must argue that the resulting diagram actually *is* the diagram of the unknot.)

Let K be a knot. Using your argument in Exercise 1.2.4, we can show that, given any diagram D for K, some number of crossings can be changed in D to produce a diagram of the unknot. Note that changing a crossing in a knot diagram is like passing the corresponding knot through itself in space. Typically, this "passing through" operation results in a different type of knot. The minimum number of times the knot must pass through itself before it becomes unknotted is called the **unknotting number** of the knot.

Note that the definition of unknotting number is a spatial characteristic of the physical knot. Thus, the unknotting number is not tied to a particular diagram of the knot. We can use diagrams to build intuition about what a knot's unknotting number might be, but such intuition does not always constitute a proof that a given knot has a specific unknotting number. For instance, if we can show that changing two crossings in a diagram of a knot K produces the unknot, we can say that the unknotting number of K is *at most* two. We cannot, however, claim that the unknotting number of K is *exactly* two without more work. Perhaps there's another diagram of the knot that only requires one crossing change to become unknotted.

Exercise 1.2.5. Use the diagrams provided to conjecture the unknotting number of the knots labeled (b), (f), and (o) from Table 1.1. Experiment with alternate diagrams for these knots to give additional support for your conjectures.

We will take a closer look at the unknotting number and study interesting ways to unknot diagrams in Chapter 7.

1.3   Alternating Knots

A diagram D for a link L is called an **alternating diagram** provided that, if you traverse each component in the diagram, then you alternately pass over and under the crossings. For example, the diagram (b) from Table 1.1 is alternating, while the diagram (l) is not. A link L is called an

alternating link provided that there exists an alternating diagram D for L.

Exercise 1.3.1. Determine which links in Table 1.1 are alternating links. Record your findings. (Note that there are many nonalternating diagrams that can be created of a given alternating link, so just because a particular diagram is not alternating does not imply that *no diagram* of that link is alternating!)

Exercise 1.3.2. Choose one of the alternating links you found in Exercise 1.3.1 that has an alternating diagram in Table 1.1. Produce an equivalent, nonalternating diagram of this link.

Exercise 1.3.3. Consider the knot projection in Figure 1.2.2. Show that the precrossings can be resolved in such a way that the diagram becomes an alternating diagram. Investigate whether this can always be done for any knot projection. Can you find a knot projection for which there is no possible selection of crossings that will result in an alternating diagram? Record your findings, arguments, and conjectures.

The following three exercises relate to finding the crossing number of an alternating diagram.

Exercise 1.3.4. Consider the alternating diagrams in Table 1.1. For each alternating diagram, can you produce an equivalent diagram for this link that has fewer crossings? If so, is the resulting diagram alternating? Record your findings.

Exercise 1.3.5. Consider the alternating knots in Figure 1.3.1. For each alternating knot, can you produce another diagram for this knot that has fewer crossings? If so, is the resulting diagram alternating? Record your findings.

Exercise 1.3.6. Suppose K is an alternating knot. Use the exercise above to state a conjecture regarding any diagram of K that has a minimum number of crossings. Are there required properties for the diagram of K? Will a diagram of K with minimum crossings always be alternating? Create your own new examples to give further evidence in support of your conjecture.

We will revisit your conjecture and provide a proof determining the crossing number for alternating links in Chapter 6.

Figure 1.3.1: Examples of alternating knots. What is the crossing number of each knot?

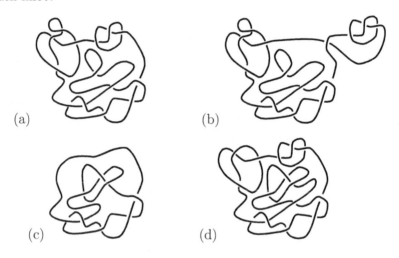

(a)   (b)

(c)   (d)

1.4   Games with Knots

Given a projection of a knot, we can play a game called the Knotting–Unknotting game. Suppose two players, Kenya and Ulysses, take the knot projection in Figure 1.4.1 as their starting "game board."

Figure 1.4.1: A game board for the Knotting–Unknotting game.

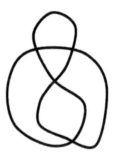

The players take turns resolving precrossings until all crossing information has been determined. Once a crossing has been resolved, it cannot be changed. **Kenya's** goal is to create a nontrivial **knot**, while Ulysses' goal is to make the **unknot**.

Exercise 1.4.1. Find a friend and decide who will play the part of Kenya and who will play the part of Ulysses. Next, decide who will play first and who will play second. Play the game on the projection in Figure 1.4.1. Who won? Did the winner seem to have an unfair advantage?

Exercise 1.4.2. Play the same game as in the previous exercise, but switch who goes first. Who won? Did the winner seem to have an unfair advantage?

Exercise 1.4.3. Now draw your own projection and play another round of the Knotting–Unknotting game. Describe any strategies you uncover for how each player should play the game.

Exercise 1.4.4. Is there a knot projection on which Ulysses (the unknotter) can always win, regardless of whether he plays first or second?

Exercise 1.4.5. Is there a knot projection on which Kenya (the knotter) can always win, regardless of whether she plays first or second?

Exercise 1.4.6. Invent and explore your own game that can be played on a knot diagram or a projection of a knot diagram.

1.5 Mirrors, Orientation & Inverses

The **mirror image** of a link diagram D, denoted D^m, is the link obtained by changing all of the crossings of D. In other words, each crossing over-strand becomes the under-strand of that crossing and vice versa.

Exercise 1.5.1. Let D be a link diagram and D^* be obtained by reflecting D through a mirror. Explain why D^* and D^m result in equivalent links. (Use the description of equivalence of knots and links that was given at the beginning of this chapter.)

A link is called **chiral** if it is *not* equivalent to its mirror image. If it is equivalent to its mirror image, it is called **amphichiral** or **achiral**.

Exercise 1.5.2. Investigate whether the knots (c), (d), (n), and (o) in Table 1.1 are chiral or achiral. Record your findings and conjectures.

A link can be given an **orientation** by simply assigning a direction of travel around each loop. Orientation is typically indicated by drawing one or more small arrowheads as shown in Figure 1.5.2.

Figure 1.5.1: A knot and its mirror image.

Figure 1.5.2: Example of an oriented knot.

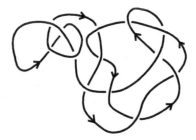

For an oriented knot diagram D, the same diagram with the opposite orientation is called the **reverse** of D, denoted $\bar{D}$. For some links, the choice of orientation does not matter. That is, the oriented link is equivalent to its reverse. A link with this property is called **invertible**.

Exercise 1.5.3. (1) Determine whether or not the links in Table 1.1 are invertible. (2) Create your own knot with seven or fewer crossings and determine whether or not your knot is invertible.

1.6   Knot Composition & Prime Knots

Given two link diagrams, we can create a new link by removing a small arc from each diagram and then connecting the four endpoints by two new arcs, as in Figure 1.6.1. For two links J and K, the new link is called the **composition** of J and K and is denoted $J\#K$.

A knot is called a **composite knot** if it can be expressed as the

Figure 1.6.1: The composition of K and L.

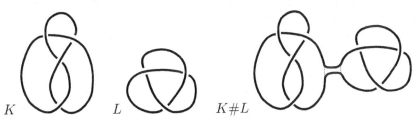

composition of two nontrivial knots. The knots that make up the composite knot are called the **factor knots**.

Exercise 1.6.1. Determine the result of composing the unknot with a link L. Record your findings.

If a knot is not the composition of any two nontrivial knots, then it is called a **prime knot**.

Exercise 1.6.2. The knots in Figure 1.6.2 are composite. Make each knot out of string. Play with your physical knots to identify their prime factor knots.

Figure 1.6.2: Examples of composite knots.

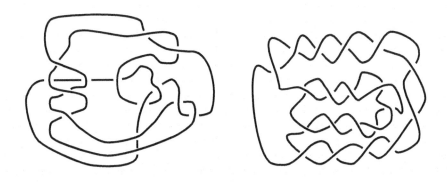

1.7   Knot Notation

Suppose you are studying an alternating knot diagram while having a phone conversation with a friend about some interesting aspect of your diagram. Suspend disbelief to further suppose that your friend doesn't have a smartphone or any other device that would allow for a picture of your diagram to be sent to her. How can you give her instructions over the phone to recreate your diagram? Is there a systematic way that you can encode or describe the diagram and crossing information so that she can reconstruct your picture?

Exercise 1.7.1. Work with a friend to create notation for encoding alternating knots that can be used in non-pictoral communication with a person or a computer. Your notation should be easy to *determine*; that is, given a knot diagram, there should be a relatively straightforward algorithm to follow to determine the notation for that particular diagram. Your notation should also be easy to *use*; that is, it should allow someone to recreate the diagram of the knot that was used to determine the notation (or recreate an equivalent diagram) with no additional information other than your notational conventions.

Exercise 1.7.2. Investigate any limitations of the notational scheme you created above by working with the same friend and following the steps provided below. Record and summarize your findings.

Step A.

1. Sit back-to-back with your friend. Each of you should have paper and pencil in hand.

2. Secretly select one of the alternating knot diagrams from Table 1.1 and use your notation to encode this knot.

3. Pass your notation only (no figures or references to Table 1.1) to your friend. They will then pass their notation to you.

4. Decode the notation given to you by drawing the diagram that the notation suggests.

5. Turn around to compare your diagram with their selected diagram from Table 1.1. Is your decoded diagram the same as or equivalent to the diagram that they encoded? Why or why not?

6. Record and summarize your findings.

Figure 1.7.1: Use this diagram in Step B.

Step B.

1. Sit back-to-back with your friend. Each of you should have paper and pencil in hand.

2. Draw the knot in Figure 1.7.1 and encode it using your notation.

3. Are there any choices made in the determination of your notation from a diagram? If so, make these choices in various ways to create a list of notations describing the diagram.

4. Exchange notations with your friend.

5. Using your friend's list of notations, draw each knot diagram.

6. Are all of the drawn diagrams equivalent? Why or why not?

7. Are there any choices you had to make when recreating the diagram from the notation? What are they? Do the various choices always result in equivalent knots? Why?

8. Repeat these steps for knot (b) from Table 1.1.

9. Record and summarize your findings.

Step C.

1. Draw your own alternating knot diagram with five or fewer crossings.

2. Encode this diagram using the notational scheme you found above.

Figure 1.7.2: A twist of an arc of the diagram.

3. Now twist one arc in your diagram so the result is a new *alternating* diagram that contains one new crossing. Figure 1.7.2 shows what is meant by the word 'twist.'

4. Encode this new diagram using your notation and swap the notation for both diagrams with your friend's notation.

5. Decode the notation given to you by your friend. Are the two diagrams equivalent?

6. Is there a recognizable characteristic within your friend's notation that could be used to identify where the twisting is encoded? Encode and decode several more examples to formulate and test your conjectures.

Exercise 1.7.3. You just designed a method for encoding alternating knots. Could your notation be modified to encode information about nonalternating knots as well?

In Chapter 4, we will learn about and investigate several ways to encode knotting information.

1.8   Questions in Knot Theory

Some of the most difficult questions in knot theory are very easy to state. Here are a few questions for us to ponder regarding the basic ideas from this chapter.

1. Given knots K and J, how can we determine if K is equivalent to J?

2. How might we prove that two knots are not equivalent? In particular, how can we prove that a given nontrivial knot is actually nontrivial?

3. How can we determine the crossing number of a knot?

4. How can we determine the unknotting number of a knot?

5. How can one produce a table of the simplest knots? How should the complexity of a knot be measured?

6. Which knots are amphichiral?

7. Given a knot K, how can we determine if K is prime?

8. Is there a definition of an alternating knot that relies only on a knot's position in space and not on any diagrams of the knot?

9. Which knots are invertible?

Exercise 1.8.1. Add your own questions to the list above.

Chapter 2

Definition and Equivalence of Knots and Links

So far, we've been exploring knot theory using intuitive ideas about what knots and links are and when they are equivalent. In this chapter, we will formalize these ideas. We begin with a definition of knots and links in $\mathbb{R}^3$ using collections of line segments.

2.1   Polygonal Curves & Δ-Equivalence

Definition 2.1.1. If A, B are points in $\mathbb{R}^3$, we use AB to denote the line segment from A to B. For an ordered set of distinct points, $(A_1, A_2, A_3, \ldots, A_n)$, with $A_i \in \mathbb{R}^3$ for all i, the union of the segments $A_1A_2, A_2A_3, \ldots A_{n-1}A_n, A_nA_1$ is called a **closed polygonal curve** in $\mathbb{R}^3$. If each segment intersects exactly two other segments, intersecting only at an endpoint, then the curve is called **simple**. We call the points A_i the **defining points** or **vertices** of the polygonal curve.

Definition 2.1.2. A **knot** is a simple closed polygonal curve in $\mathbb{R}^3$. A **link**, L, is a finite union of pairwise nonintersecting knots. The number of knots in the union comprising L is called the number of **components** of the link. Thus a knot is a link with exactly one component.

Figure 2.1.1: Example of a knot.

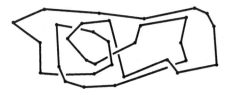

Figure 2.1.2: Example of a link with three components.

Definition 2.1.3. The two-dimensional pictures of links, like those seen in Figures 2.1.1 and 2.1.2, are called **link diagrams**. A link diagram comes

from a *particular type* of **projection** (in the standard geometric sense) of the link onto a plane in $\mathbb{R}^3$. At each point where the projection of two line segments from L results in a point of intersection in the projection plane, one of the two segments is drawn as broken, or as if a piece of the line segment is missing. The broken segment is assumed to be passing under the unbroken segment. The places in a diagram where a segment passes under another is called a **crossing** of the diagram.

The projection of a link L onto two distinct planes in $\mathbb{R}^3$ can result in strikingly different images, and some projections cannot be used to recreate a knot in space unless more information is provided. Such projections should be avoided for the purposes of our discussion. In the exercise below, we investigate projections and create a succinct list of properties to avoid.

Exercise 2.1.4. Make a knot out of straws and tape with 9 straws. Complete the four steps below.

1. Draw a projection of your knot that looks as if only 8 straws were used in the knot.

2. For the same knot, draw several projections from which the knot cannot be reconstructed by simply indicating crossing information. Draw any other projections you encounter that contain diagrammatic oddities.

3. Make a list of all properties of projections that should be avoided when drawing a link diagram.

4. Compare your list with a friend's list to create a complete and succinct list of Properties to Avoid.

Definition 2.1.5. A **regular projection** of a link L is a projection that has none of the Properties to Avoid from Exercise 2.1.4, part 4. A **link diagram** for L is a regular projection of the link L that includes all crossing information (i.e., information about which strand passes over and which passes under at each crossing).

It is not immediately obvious that all links have regular projections. Sometimes a link must be manipulated in $\mathbb{R}^3$ before a regular projection can be found. However, given any link L, there will always exist a link diagram that represents either L or a link equivalent to L.

Definition 2.1.6. Let A and B be two adjacent defining points of a link L in $\mathbb{R}^3$. Suppose C is a point in $\mathbb{R}^3$ such that the triangle $\triangle ABC$ and its interior intersect L only along the line segment AB. We call such a triangle, $\triangle ABC$, an **elementary triangle**. An elementary triangle can be a **degenerate triangle** with $C \in AB$.

Under these conditions, replacing the line segment AB by the segments $AC \cup CB$ results in a new link, L'. This process of replacement is called an **elementary move** applied to L resulting in L'. The inverse process of removing $AC \cup CB$ and replacing these two segments with the segment AB, assuming that the triangle $\triangle ABC$ and its interior intersect the link only along the segments $AC \cup CB$, will also be referred to as an elementary move.

When applying an elementary move $\triangle ABC$ to a diagram of a link, we always select a point C that both satisfies the definition of an elementary triangle and results in a link diagram for L'. For if a point $C \in \mathbb{R}^3$ is selected such that the resulting diagram is not regular, then the projection plane or the point C can be altered slightly to result in a projection to a link diagram.

Figure 2.1.3: Example of an elementary move applied to L resulting in L'.

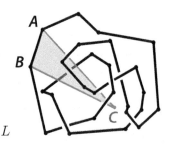
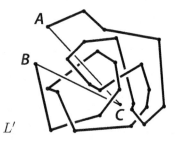

Definition 2.1.7. Two links, L and L', are called **Δ-equivalent** or **delta equivalent**, denoted $L \overset{\Delta}{\sim} L'$, provided there exists a sequence of links $L = L_1, L_2, \ldots, L_n = L'$ such that each L_{i+1} is obtained from L_i via an elementary move, for all $i = 1, \ldots n - 1$.

Exercise 2.1.8. Prove that Δ-equivalence is an equivalence relation on the set of links. In other words, prove that $\overset{\Delta}{\sim}$ is a reflexive, symmetric, and transitive relation.

Exercise 2.1.9. Prove that the knots K_1 and K_2, shown in Figure 2.1.4, are Δ-equivalent. (See Figure 2.1.5 for examples of elementary moves applied to K_1.)

Figure 2.1.4: Example of equivalent knots.

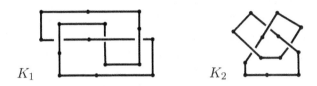

Figure 2.1.5: A sequence of elementary moves applied to K_1. The red segments outline the elementary triangle and indicate the new line segment(s) that join the diagram after the elementary move is applied. Each arrow, except for one, is an equivalence resulting from a single elementary move. Can you spot the arrow that is the result of two elementary moves?

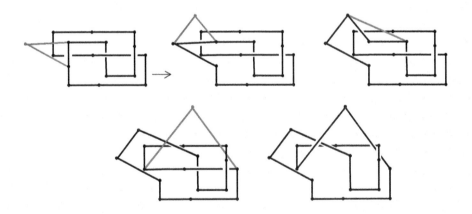

Exercise 2.1.10. Determine which equivalences shown in Figure 2.1.5 are the result of more than one elementary move.

2.2   Diagram Equivalence via R-Moves

As we saw in Exercise 2.1.9, creating and visually following a sequence of elementary moves is not always easy. In this section, we develop an alternate but equivalent notion of link equivalence that is generated by a small list of localized changes to a link diagram. To find this list of small changes, we consider subdivisions of elementary triangles into 'less complicated' elementary triangles. To make this more concrete, let's look at an example. Consider the elementary triangle in (a) from Figure 2.2.1 and the subdivision shown in (b). Each subtriangle in the subdivision contains *at most two line segments* of the link diagram. If we can show that for *any* elementary triangle such a subdivision can always be found, then we can redefine link equivalence in terms of a sequence of elementary moves using only our 'less complicated' elementary triangles from a small finite list. This reframing of link equivalence will prove to be very powerful in Chapter 5 where we study combinatorial properties of links.

Figure 2.2.1: An elementary triangle $\triangle ABC$ and a subdivision. Each subtriangle of the subdivision is itself an elementary triangle. The elementary move $\triangle ABC$ can also be performed by applying all the elementary moves in the subdivision one at a time, starting along the segment AB and moving upward toward C. (Check this! Several of the elementary moves must use degenerate triangles.)

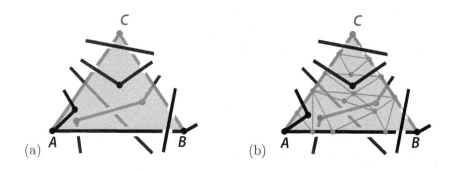

Our goal is to create a short list of simple, localized, elementary moves that is long enough so that *any* elementary triangle can be subdivided into pieces on our list. Do you see the tension in this goal? A *short* list that is *long* enough. In this section, we look to Figure 2.2.1 for inspiration of what to put on our list of moves, but there is no guarantee that this example subdivision contains all the moves that we will need. (In fact, it does not!) Once our list is seemingly complete, then we must *prove* that our list is long enough so that any elementary triangle can be subdivided into moves on our list.

The first additions to our list of localized elementary moves are a collection of simple moves that do not change the number of crossings in a link diagram. Examples of these moves, called planar isotopies, are shown in Figure 2.2.2.

Definition 2.2.1. Let $\triangle ABC$ be an elementary triangle in a diagram D that contains the link segment AB. The elementary move $\triangle ABC$ is called a **planar isotopy** if it satisfies one of the following three conditions.

(i) *In the link diagram, $\triangle ABC$ coincides with no points of the link other than the segment AB.*

(ii) *In the link diagram, $\triangle ABC$ and the segment AB lie under (or over) the interior of exactly one line segment that is not adjacent to AB. If the line segment crosses under the segment AB (resp., over AB), then it crosses under (resp., over) either AC or CB after replacement.*

(iii) *In the link diagram, $\triangle ABC$ and the segment AB lie under (or over) exactly two adjacent line segments and their shared vertex. One of the two segments crosses over (resp., under) the segment AB, thus the other segment crosses over (resp., under) either AC or CB after replacement.*

The inverse of a move of type (i)-(iii) is also a planar isotopy. A sequence of several planar isotopies is also considered to be a planar isotopy.

The dotted circle in Figure 2.2.2 indicates that the move is taking place locally, i.e., within a small region of the diagram and the diagram is unchanged outside this region.

Figure 2.2.2: Planar isotopies of types (i), (ii), and (iii).

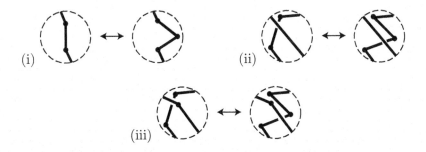

Exercise 2.2.2. Number all the subtriangles in Figure 2.2.1 (b) to indicate an order in which they can be applied. (There are many valid ways to do this.) For your ordering, determine the number of subtriangles that are planar isotopies of type (i), determine how many are of type (ii), and find how many are of type (iii). Check with a friend to see if your numbers are the same. Your numbers might be different depending on the order in which the subtriangle moves are made.

Exercise 2.2.3. Let L_1 and L_2 be the links in Figure 2.2.3. Provide a sequence of planar isotopies from L_1 to L_2.

Figure 2.2.3: The link L_2 can be obtained from link L_1 via a sequence of planar isotopies.

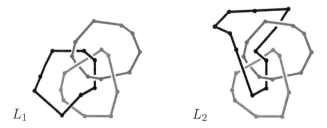

L_1 L_2

Several of the subtriangles in Figure 2.2.1 (b) are not planar isotopies because they change the number of crossings in the diagram. Therefore, we must add to our list of localized elementary moves a few simple moves that *do change* the number of crossings. Figure 2.2.4 contains two examples of elementary moves that increase or decrease the number of crossings in the diagram by one. We add these elementary moves to our list.

Figure 2.2.4: This elementary move, called an R1 move, changes the number of crossings in the diagram by one.

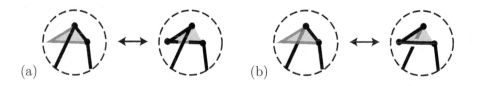

(a)   (b)

Definition 2.2.4. Let $\triangle ABC$ be an elementary triangle that lies over or under a single edge and that edge is adjacent to an edge that is to be replaced (as is the case in Figure 2.2.4). Then the elementary move $\triangle ABC$ is called an **R1 move**.

Exercise 2.2.5. Find at least one example of an R1 move subtriangle in the subdivision shown in (b) of Figure 2.2.1.

Before adding more moves to our list, let's eliminate from consideration a special type of elementary triangle that an R1 move is designed to address. The next two lemmas will help when proving that our list is "long enough" because they allow us to focus our proof on certain "nice" elementary triangles.

From the definition of a link, there are two line segments that emanate from a segment AB, and there are two possibilities of how an elementary triangle might interact with the two line segments of the diagram that emanate from AB. The elementary triangle $\triangle ABC$ either (1) doesn't coincide with the edges emanating from AB; or (2) coincides with at least one of the edges emanating from AB. The simpler situation of type (1) is pictured in Figure 2.2.5, where neither of the edges emanating from AB lies above and neither lies below the elementary triangle ABC. Two examples of type (2) can be found in Figure 2.2.6.

Figure 2.2.5: Elementary move of type (1). The edges emanating from AB do not coincide with triangle $\triangle ABC$.

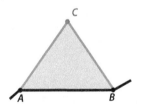

In the lemma below, we will show that for an elementary triangle of type (2) where exactly one edge emanating from AB coincides with the elementary triangle, it is possible to subdivide the elementary triangle into a sequence of three elementary moves, two of which are type (1) and the other is an R1 move. This lemma will allow us to concentrate our efforts on elementary triangles of type (1) when looking for any remaining necessary additions to our list.

Lemma 2.2.6. *Suppose A and B are adjacent defining points of a link L with diagram D. Suppose $C \in \mathbb{R}^3$ is a point such that the triangle $\triangle ABC$*

is an elementary triangle of type (2) with exactly one emanating edge coincident with △ABC. Then the elementary move corresponding to the triangle ABC can be performed as a composition of three elementary moves: an R1 move and two elementary moves of type (1).

Figure 2.2.6: Elementary moves of type (2). On the left, one of the edges emanating from AB lies under the elementary triangle. On the right, the elementary triangle coincides with both edges emanating from AB.

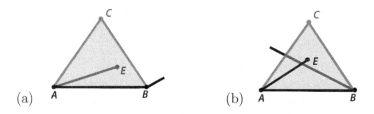

Figure 2.2.7: An elementary triangle that coincides with one of the edges emanating from AB as well as other portions of the diagram D.

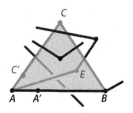

Proof of Lemma 2.2.6. Let EA denote the line segment that lies above or below the elementary triangle ABC. Select a point A' along the segment AB so that (i) the segment $A'C$ doesn't intersect any vertices or crossings in the diagram D, (ii) the triangle $AA'C$ contains no vertices or crossings from D, and (iii) the segment AA' does not contain a crossing. Such an A' exists since there are only finitely many vertices and crossings in D, but there are infinitely many candidates for A' along the segment AB. Notice that the triangles $AA'C$ and $A'BC$ are both elementary triangles since they are subtriangles of $△ABC$.

Next, select a point C' along the segment AC such that the triangle $AA'C'$ coincides with no parts of the link diagram other than the edge AA' and a portion of the edge AE. In particular, select C' so that the segment AC'

contains no crossings. Why does such a point C' exist? If a point C' were selected and there were some other part of the diagram D that coincided with $\triangle AA'C'$, then, by condition (ii) in the selection of A', the incident portion of D would be a straight edge that crosses $\triangle AA'C$. Where might this edge cross $\triangle AA'C$? It could not cross the segment AA' by condition (iii) in the selection properties of A'. Hence, any straight edge that coincides with $\triangle AA'C'$ must intersect both of the edges AC' and $C'A'$. Since A' was selected to satisfy (i), this edge could not cross AE within $\triangle AA'C'$. Hence the edge must coincide with $\triangle AA'C'$ on the side of AE opposite AA'. In this case, we make a new selection for C' that is close enough to A so that AC' contains no crossings of D. Then the triangle $AA'C'$ will not coincide with the diagram D except along AE.

To complete the proof of Lemma 2.2.6 we perform the elementary move $\triangle ABC$ via a sequence of three subelementary moves. (See Figure 2.2.8 for reference.) First apply the move that replaces segment AA' with segments $AC' \cup C'A'$ (corresponding to $\triangle AA'C'$), which, by design, is an R1 move. Next apply the moves $\triangle A'C'C$ and $\triangle A'CB$ which, by construction, are both of type (1). ∎

Figure 2.2.8: A sequence of three elementary moves that results in the move $\triangle ABC$ when composed. Starting with the elementary triangle in Figure 2.2.7, first apply the R1 move $\triangle AA'C'$, as seen in (i). Then apply $\triangle A'C'C$, as seen in (ii), and $\triangle A'CB$, as seen in (iii).

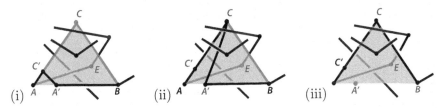

The previous lemma addresses elementary triangles of type (2), where *exactly one* edge emanating from A or B lies above or below $\triangle ABC$. How do we deal with type (2) triangles where *both* emanating edges lie above or below $\triangle ABC$? We consider this subcase in the following lemma.

Lemma 2.2.7. *Suppose A and B are adjacent defining points of a link L and $C \in \mathbb{R}^3$ is a point such that the triangle $\triangle ABC$ is an elementary triangle. Let D be a diagram for $L \cup \triangle ABC$. Suppose that both edges adjacent to AB lie above or below the elementary triangle in D. Then the*

elementary move $\triangle ABC$ *can be subdivided into a composition of*
elementary moves: two R1 moves and other elementary moves of the type
in case (1).

Exercise 2.2.8. Use Lemma 2.2.6 to prove Lemma 2.2.7.

Our list of localized elementary moves currently contains planar isotopies
of types (i), (ii), and (iii) and the R1 move, but, as seen in Figure 2.2.1,
there are still more moves we must add. Thanks to the previous two
lemmas we can reduce our search from considering all possible elementary
triangles to now only considering elementary triangles of type (1). This
makes our search a bit more manageable.

Using Figure 2.2.5 as a starting point, let's systematically consider how
pieces of diagram D may lie above or below $\triangle ABC$. Consider the cases
where

(a) no part of link L lies above or below $\triangle ABC$;

(b) exactly one line segment lies above or below $\triangle ABC$;

(c) exactly two line segments lie above or below $\triangle ABC$; or

(d) more than two line segments lie above or below $\triangle ABC$.

Clearly these cases exhaust all possibilities. Case (a) describes a planar
isotopy of type (i), which is an elementary triangle that has already been
added to our list.

Exercise 2.2.9. For cases (b) and (c) listed above, draw an exhaustive list
of representatives (up to planar isotopy) of how the line segments may be
situated relative to triangle $\triangle ABC$. For example, in case (b) the line
segment could cross AB or not cross AB. The subcase where it crosses AB
has already been considered. Explain why. In the subcase where the
segment doesn't cross AB there are two cases, as shown in Figure 2.2.9.
Case (c) can be separated into three subcases: the two line segments do
not intersect above or below $\triangle ABC$; the two line segments are adjacent
and thus $\triangle ABC$ contains exactly one defining point; or the two segments
cross within $\triangle ABC$. Note that, up to planar isotopy, we don't care where
a given line segment crosses AB, just that it does.

Exercise 2.2.10. First, determine which of the figures in your solution to
Exercise 2.2.9 result in an elementary move that is a planar isotopy. These

Figure 2.2.9: Examples of case (b).

moves are already on our list. Next, determine which of your figures are merely mirror images of each other. List only one figure from each mirror image pair and add it to our list.

In an effort to keep our list of elementary moves short, we will eliminate a few moves. If an elementary move on our list can be completed using other elementary moves already on our list, then we need not include it on our list. Such moves are called **redundant**.

Exercise 2.2.11. Prove that the move in Figure 2.2.10 is redundant. That is, show that this move can be completed through a sequence of other moves on our list. (Hint: Consider the subdivision in Figure 2.2.10.)

Figure 2.2.10: This elementary move can be subdivided to show that it is redundant. Are any additional subdivisions needed?

Exercise 2.2.12. Continuing with your list from Exercise 2.2.10, remove all moves that are redundant. Provide a proof that each removed move is actually redundant.

Now we should have a complete list of all necessary R-moves up to mirror images. Check your list with Figure 2.2.11.

Definition 2.2.13. The moves listed in Figure 2.2.11 are called **R-moves**. Link diagrams that are equivalent through a sequence of R-moves and planar isotopies are called **R-equivalent**.

Figure 2.2.11: Is this list of R-moves complete?

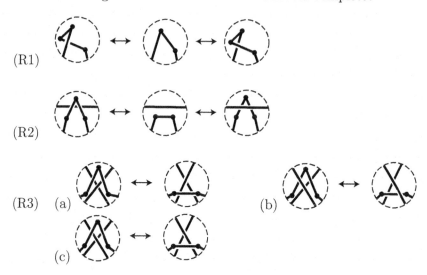

(R1)

(R2)

(R3)   (a)   (b)

       (c)

Last, we consider case (d), where more than two line segments of L lie above or below the type (1) elementary triangle $\triangle ABC$. For this case, we subdivide the elementary triangle and show that our existing list of R-moves is sufficient.

Proposition 2.2.14. *Suppose A and B are defining points of a link L with diagram D and the triangle $\triangle ABC$ is an elementary triangle for L such that the edges emanating from AB do not lie above or below $\triangle ABC$ in diagram D. Suppose that more than two line segments of L lie above or below $\triangle ABC$. Then $\triangle ABC$ can be subdivided into elementary subtriangles such that no more than two line segments of L lie above or below each subtriangle.*

Exercise 2.2.15. Prove Proposition 2.2.14 by describing an algorithm that results in a subdivision of $\triangle ABC$ with the desired properties. (Hint: Start by drawing an example triangle and place several segments of L above and below the example triangle. Find a *process* that can be used algorithmically to subdivide your example triangle into subtriangles. Then think about how to describe and generalize your algorithm so it may be applied to an arbitrary $\triangle ABC$. In your proof, make sure to address why your algorithm will *always* work.)

The results of this section can now be used to prove the main result of the next section, that Δ-equivalence is equivalent to R-equivalence.

2.3   The Equivalence of Δ- and R-Equivalence

Theorem 2.3.1. *Let L and L′ be links and D and D′ be diagrams of L and L′, respectively. Then L and L′ are Δ-equivalent if and only if D and D′ are R-equivalent.*

Exercise 2.3.2. Prove Theorem 2.3.1.

Figure 2.3.1: The three types of Reidemeister moves.

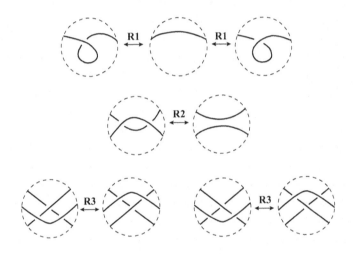

Knots and links are usually thought of and drawn as smooth curves. An informal way of dealing with this is to view the smooth arcs as polygonal curves with a very large number of segments arranged so that the curve appears smooth. From here onward, we will draw knots and links as smooth. The smoothed versions of the R-moves, shown in Figure 2.3.1, are merely the R-moves along with many planar isotopies, enough to make the diagram look smooth. We will often refer to these smoothed moves as *Reidemeister moves*, in honor of Kurt Reidemeister, the mathematician that discovered them [31]. We consider Reidemeister moves to be smooth R-moves together with smooth planar isotopies. If we prefer to be brief, we may also use R-moves to refer to the Reidemeister moves.

Definition 2.3.3. The three **Reidemeister moves**, often denoted R1, R2, and R3, are given pictorially in Figure 2.3.1. We assume that when a Reidemeister move is performed on a link diagram, the part of the diagram located outside of the dotted circle remains unchanged.

Figure 2.3.2: Two smooth knots that are equivalent via Reidemeister moves.

Exercise 2.3.4. The knots in Figure 2.3.2 are equivalent via Reidemeister moves. An equivalence is started in Figure 2.3.3, where each arrow denotes the application of a single Reidemeister move. Complete the proof of this equivalence. Each diagram in your equivalence should differ from the previous diagram by no more than *one* Reidemeister move and/or a planar isotopy.

Notice in Figure 2.3.2, the equivalence from (b) to (c) denotes an R3 move. An R3 move locally transforms a diagram with three crossings and removes no crossings. Do you see why the loop in (c) is a required outcome of the R3 move?

Figure 2.3.3: A sequence of Reidemeister moves applied to K_1.

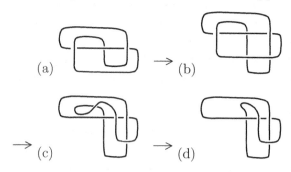

Exercise 2.3.5. Consider the knot projection P shown in Figure 2.3.4. Surprisingly, for each knot K in Figure 2.3.5, there is some choice of crossing information of P that yields K. (Indeed, the projection in

Figure 2.3.5 contains *all knots* that can be drawn with six or fewer crossings!) So for each knot K in Figure 2.3.5, your task is to choose crossing information for the precrossings in P to obtain K. Then use Reidemeister moves to show that your resolution of crossings in P yields a knot diagram of K equivalent to the particular diagram of K given in Figure 2.3.5.

Figure 2.3.4: A 7-crossing knot projection, P.

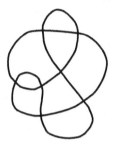

Figure 2.3.5: All knots that can be drawn with 6 or fewer crossings.

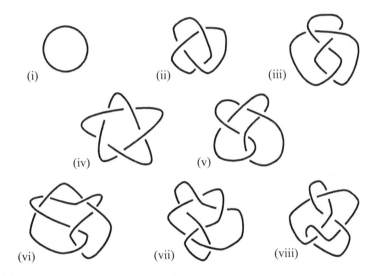

Exercise 2.3.6. In Figure 2.3.1, we illustrate two types of R1 moves, one R2 move, and two types of R3 moves, for a total of five oriented moves. If we were to draw all possible *oriented* Reidemeister moves, i.e., R-moves where we choose an orientation for each strand involved in the move, how many moves would we generate?

Mathematician Michael Polyak proved that, from our large set of oriented R-moves, there is a subset of just four R-moves that generate the entire set [30]. What exactly does this mean? Just as any vector in a vector space can be written as a linear combination of basis vectors, any oriented R-move can be derived from a sequence of R-moves from our generating set. It turns out there are many generating sets of four R-moves. We give one particular generating set in Figure 2.3.6.

Figure 2.3.6: A generating set of oriented Reidemeister moves.

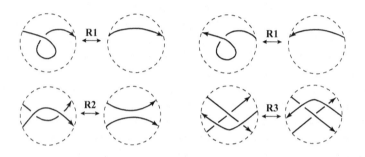

Exercise 2.3.7. In Figure 2.3.7, we picture an oriented R2 move that is not in the generating set shown in Figure 2.3.6. Show how this R2 move can be derived from a sequence of moves in the generating set.

Figure 2.3.7: An oriented R2 move.

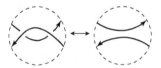

2.4 Nonequivalence and Invariants

Now that we have a well-defined notion of equivalence for links and link diagrams, we can begin to explore the idea of *nonequivalence*. To show that two diagrams represent equivalent knots or links, we need only find a sequence of Reidemeister moves that transforms one into the other. But how is it possible to show that two diagrams *fail* to be equivalent? If you think about it, you can see that failing to find a Reidemeister sequence relating two link diagrams is not *proof* that the two diagrams represent different links. Indeed, there are some complicated diagrams of the unknot that need to be made more complicated before they can be simplified. Sometimes a great deal of creativity is required to find a Reidemeister sequence relating two diagrams.

The answer to the riddle of how to prove nonequivalence is contained in the notion of an *invariant*. Informally, a **link** or **knot invariant** is a function that takes in a link or knot and outputs a value. If two equivalent links or knots are plugged into this function, the function *must* return the same or equivalent values. Conversely, if two nonequivalent links or knots are plugged in, the invariant may return the same or different values. In the case when two links are plugged in and two distinct values are returned, we can be sure that the two input links are distinct. So link invariants provide tools to prove that certain link diagrams are nonequivalent.

As an analogy, suppose you encounter a strange character (call him Mr. F) who you suspect to be your friend Tobias in disguise. To determine if Mr. F is indeed Tobias, you might apply the "height invariant." Height is typically a fixed quality in young and middle aged adults. So if you determine that Tobias is about 6'1" and Mr. F is 5'7", you *can be sure* that Mr. F is not Tobias in disguise.

Now that we have some intuition, let us introduce the formal definition of an invariant.

Definition 2.4.1. A **knot** or **link invariant** is a function from the set of equivalence classes of knots or links to a specified codomain.

In Chapter 5 and beyond, we will study knot and link invariants that are functions into the set of integers, a set of matrices, or a set of polynomials. To prove that a function is indeed a link invariant, we must simply show that it is a well-defined function on the set of equivalence classes of links. In other words, we must show that if a link K is equivalent to a link J, then the function has the same value on K as is does on J.

Before we develop invariants, however, let's learn about some interesting examples of knots and links!

Chapter 3

Families of Links and Braids

Whenever you learn a new mathematical concept, it is useful to have a collection of simple examples to apply the new concept to in order to gain intuition. For instance, $\mathbb{R}^2$ is a nice example of a vector space, $\mathbb{Z}$ is a useful example of a group, and the torus is an interesting, yet simple, example of a topological space. In this chapter, we introduce a few families of knots and links with particularly nice properties. These examples will provide some context for us to think about the knot invariants we will introduce in Chapter 5.

3.1   Twist Knots

Twist knots are one of the simplest families of knots. They will be an interesting family to consider when we look at various combinatorial properties of knots.

Definition 3.1.1. A **twist knot** of n half-twists, denoted T_n, is obtained by twisting two parallel strands n times and then hooking the ends together so that the knot is alternating, as seen in Figure 3.1.1.

Figure 3.1.1: Examples of twist knots.

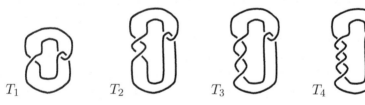

T_1 $\qquad$ T_2 $\qquad$ T_3 $\qquad$ T_4

Exercise 3.1.2. Prove that knots (b) and (c) from Table 1.1 are twist knots. Are there any other twist knots in Table 1.1?

Exercise 3.1.3. Prove that every twist knot is invertible. (Recall the definition of invertible from Section 1.5.)

Exercise 3.1.4. Investigate the Knotting-Unknotting game (from Section 1.4) on the projection of twist knots T_n where n is even. Prove the following propositions.

1. If Kenya (the knotter) plays second on a twist knot T_n, where n is even, then Kenya has a winning strategy.

2. If Ulysses (the unknotter) plays second on a twist knot T_n, where n is even, then Ulysses has a winning strategy.

Exercise 3.1.5. Investigate the Knotting-Unknotting game on the projection of twist knots T_n where n is odd. Who has a winning strategy when Kenya (the knotter) plays second? What about when Ulysses (the unknotter) plays second? Formulate and prove two propositions about these cases.

3.2 Pretzel Links

Another fascinating collection of links is the family of pretzel links. These links are both rich in structure but also easy to visualize since they are made out of simple twists.

Definition 3.2.1. Let p, q, and r be integers. A **3-strand pretzel link** $P_{p,q,r}$ can be constructed as follows. Take three pairs of string segments and arrange them vertically. Twist the bottom ends of the first pair p times (in the counterclockwise direction if $p > 0$ and in the clockwise direction if $p < 0$). Twist the bottom ends of the second pair q times and the bottom end of the third pair r times. After twisting the pairs, glue (or tie) more strands of string to adjoin the ends of the three pairs as in the examples pictured in Figures 3.2.1 and 3.2.2.

Figure 3.2.1: The pretzel link $P_{5,-3,7}$.

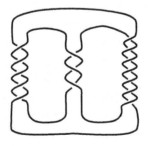

Exercise 3.2.2. How many components do the pretzel links in Figures 3.2.1 and 3.2.2 have? Is either one of these links a knot?

Figure 3.2.2: The pretzel link $P_{4,-2,-5}$.

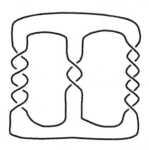

Exercise 3.2.3. For which values of p, q, and r is the link $P_{p,q,r}$ a knot? For which p, q, and r does $P_{p,q,r}$ have more than one component? Conjecture conditions on p, q, and r that determine the number of components in $P_{p,q,r}$ and then prove your conjecture.

Exercise 3.2.4. (i) Investigate the invertibility of the knot $P_{7,5,4}$.
(ii) Investigate the invertibility of the knot $P_{7,5,3}$.

Exercise 3.2.5. (i) Prove that the pretzel knot $P_{5,-1,-1}$ is equivalent to the twist knot T_4.
(ii) In general, are there values of p, q, and r such that $P_{p,q,r}$ is equivalent to T_n?

Exercise 3.2.6. Investigate if there is a relationship between $P_{a,b,c}$ and $P_{b,c,a}$ and $P_{c,a,b}$. If you find a relationship, explain it. Furthermore, explore if there is a relationship between $P_{a,b,c}$ and a pretzel link with a noncyclic permutation of the original link's indices, like $P_{a,c,b}$. What did you discover?

Exercise 3.2.7. Determine a condition on the integers a, b, and c that guarantees $P_{a,b,c}$ is alternating.

Exercise 3.2.8. Prove that for positive integers q and r, the link $P_{q,-1,r}$ is an alternating link.

3.3   Torus Links

Torus links are one of the most accessible, symmetric, and interesting infinite families of links. We see several examples of knots and links in this family in Figure 3.3.1.

Figure 3.3.1: Examples of torus links.

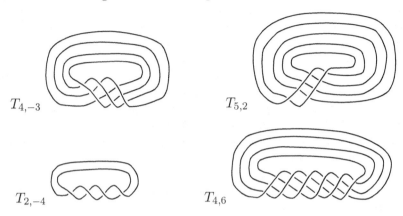

These links are called 'torus links' because each member of this family can be embedded on the surface of a torus (think: doughnut) without any crossings, as shown in Figure 3.3.2.

Figure 3.3.2: The torus link $T_{3,-2}$ embedded on the surface of a torus.

Definition 3.3.1. A **torus link diagram** $T_{p,q}$ is defined if $p > 0$ and $p, q \in \mathbb{Z}$. Begin with p horizontal strands. For q positive, the bottommost q strands are—one at a time—wrapped up and over the remaining strands. For q negative, the topmost q strands are—one at a time—wrapped down and over the remaining strands. Once the wrapping is complete, the strand ends are connected so that no new crossings are made. A link L is called a **torus link** if it is equivalent to a torus link diagram.

Exercise 3.3.2. Draw torus links $T_{6,9}$ and $T_{10,-6}$. For the four examples given in Figures 3.3.1, as well as links $T_{6,9}$ and $T_{10,-6}$, determine the number of components in each link. Which are knots?

Exercise 3.3.3. Conjecture a condition on p and q for $T_{p,q}$ which results in a one-component torus knot and prove your conjecture.

Exercise 3.3.4. The families of pretzel links and torus links are not distinct. Show that the pretzel knot $P_{-2,3,3}$ is a torus knot. Can you find another pretzel knot that is a torus knot?

Exercise 3.3.5. Investigate whether or not the family of twist knots is distinct from torus knots. Is there a twist knot that is also a torus knot? If so, give an example. If not, explain why not.

Exercise 3.3.6. Prove that the torus knot $T_{3,4}$ is equivalent to the knot in Figure 3.3.3, which is known as the knot 8_{19}.

Figure 3.3.3: The knot 8_{19} is a torus knot.

Exercise 3.3.7. Investigate whether or not the knot 8_{19} is an alternating knot.

Exercise 3.3.8. Investigate whether or not the knot $T_{2,3}$ is equivalent to $T_{3,2}$. Can your findings be generalized?

3.4 Closed Braids

The family of links called closed braids are a generalization of torus links. We define closed braids by considering a diagram of n concentric strands. For example, see Figure 3.4.1 with 4 concentric strands. A small disc-like subregion of the diagram is called a **replaceable region** if it contains arcs from two adjacent concentric strands and nothing else. See Figure 3.4.2 for examples and nonexamples of replaceable regions.

Definition 3.4.1. A **closed n-braid diagram** is obtained from a diagram of n concentric strands by identifying finitely many nonintersecting replaceable regions and replacing the two strands within each region by a crossing. A knot or link is a **closed braid** if it is equivalent to a closed n-braid diagram for some n.

Figure 3.4.1: Four concentric strands.

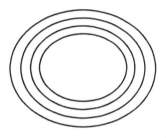

Figure 3.4.2: Examples of replaceable regions are shown in (a). The diagram (b) shows a nonexample of a replaceable region because it contains three strands.

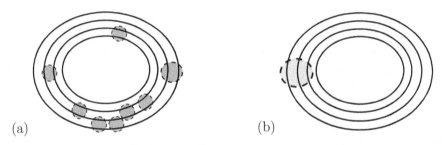

(a)   (b)

Figure 3.4.3: An example of a closed braid resulting from placing crossings in the replaceable regions shown in Figure 3.4.2 (a).

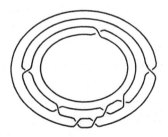

By construction, a closed n-braid diagram has an innermost region with the property that any line segment drawn from the innermost region to a point that is exterior to the link will intersect the diagram at exactly n points. We count both the under- and over-strands if the segment happens to pass through a crossing. In Figure 3.4.4, we see an example that is not in closed braid form because it fails the line-segment condition.

Figure 3.4.4: A diagram that is not in the form of a closed braid diagram.

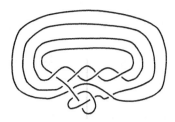

Closed braids can have one or more components. In Figure 3.4.5, the closed braid (a) is a link while the closed braid (b) is a knot.

Figure 3.4.5: (a) A two component closed braid link, and (b) a closed braid knot.

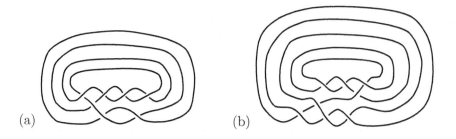

(a)   (b)

Exercise 3.4.2. Identify concentric circles and replaceable regions to show that the torus link $T_{p,q}$ is a closed braid.

As we saw in Chapter 2, diagrams that look quite different can be equivalent via Reidemeister moves and, thus, represent the same knot. In our current setting, this means that a knot is a closed braid if it *can be represented* in a closed braid diagram. Let's investigate whether certain examples of the knots and links we've already studied are closed braids. If a given example does not initially look like a closed braid, could they be changed via Reidemeister moves into a diagram of a closed braid?

Exercise 3.4.3. Identify six diagrams in Table 1.1 that, up to planar isotopy, are in closed braid form.

Exercise 3.4.4. Play with the knot in Figure 3.4.6 to discover how to put it into closed braid form. (Hint: Make the region with the dot the innermost region. To do so, only one strand needs to be moved—the bold strand—but the movement will require several Reidemeister moves. Write down a sequence of diagrams that show how to put the knot into closed braid form.)

Figure 3.4.6: Is this knot diagram equivalent to a closed braid diagram?

Exercise 3.4.5. Show that knot (n) from Table 1.1 can be written in closed braid form.

Exercises 3.4.6 and 3.4.5 show that even though a knot diagram may seem far from being in closed braid form, it is sometimes possible to manipulate it into a closed braid form. Is this always possible? Surprisingly, the answer is yes! In preparation to prove that all knots are closed braids, we need some new notation that uses the sign of a crossing.

Definition 3.4.6. For a crossing in an oriented link diagram, the **sign** of the crossing is determined by the "right hand rule.'" Imagine placing your thumb on the over-strand of the crossing, pointing in the direction of the orientation, and curling your fingers under and around in the direction of the under-strand. A crossing is called a **positive crossing** when you *must* use your right hand to curl your fingers in the direction of the under-strand. A crossing is called a **negative crossing** when you *must* use your left hand to curl your fingers in the direction of the under-strand. Figure 3.4.7 shows examples of positive and negative crossings.

Figure 3.4.7: A positive crossing and a negative crossing.

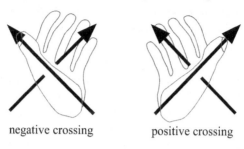

negative crossing   positive crossing

We introduce an alternative diagrammatic notation that highlights certain features of a closed braid. Here is how it works for diagrams that are already in closed braid form. Start by giving the closed braid diagram an orientation such that all concentric strands will be traversed clockwise or all will be traversed counterclockwise about the innermost region. In the alternative notation for our braid diagram we will replace each positive crossing of the braid with a line segment labeled with a positive sign, "+," and replace each negative crossing with a line segment labeled with a negative sign, "–". Figure 3.4.8 shows an example of how this is done for a diagram in closed braid form.

Figure 3.4.8: An example of an oriented closed braid and the alternate notation for this braid consisting of oriented circles and signed-line segments.

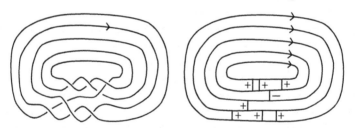

The alternate notation described above is easy to turn back into a closed braid. As seen in Figure 3.4.8, we simply replace each signed-line segment with a crossing of the corresponding sign.

To prove that *every* knot is equivalent to a closed braid, we develop an analogous way to transform an arbitrary knot diagram into a diagram of oriented circles and signed-line segments. As is the case for the alternate notation applied to a closed braid diagram, we will create oriented circles that *consistently follow the orientation of the diagram* and signed-line segments that *connect* the oriented circles. Thus, given an arbitrary knot diagram, we will smooth each crossing so that the local orientation of the arcs are consistent with the orientation of the diagram. Then, we replace the crossing with an appropriately signed-line segment connecting the two oriented arcs. Each newly formed circle in the resulting diagram has a well-defined orientation because each local smoothing respected the global orientation of the knot. An example of this process applied to a negative crossing is shown in Figure 3.4.9 and an example of it applied to a positive crossing is shown in Figure 3.4.10.

Figure 3.4.9: (a) An arbitrary negative crossing, (b) a smoothing of the crossing that is consistent with the orientation of the diagram, and (c) a replacement of the crossing with a signed-line segment that connects the two arcs.

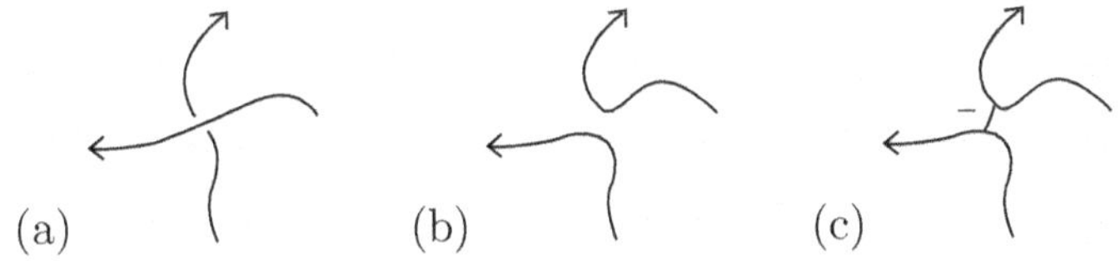

Figure 3.4.10: (a) An arbitrary positive crossing, (b) a smoothing of the crossing that is consistent with the orientation of the diagram, and (c) a replacement of the crossing with a signed-line segment that connects the two arcs.

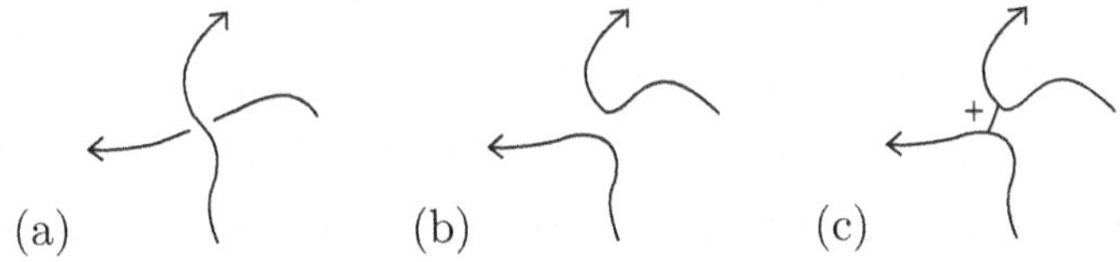

Definition 3.4.7. The **Seifert diagram** of an oriented knot diagram is the diagram of oriented circles and signed-line segments that is obtained by applying the smoothing and replacement procedure, shown in Figures 3.4.10 and 3.4.9, to every crossing in the oriented diagram. The orientated circles in the Seifert diagram are called **Seifert circles**.

The twist knot T_4 and its Seifert diagram are shown in Figure 3.4.11.

Figure 3.4.11: The knot T_4 and its Seifert diagram.

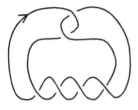 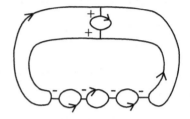

Exercise 3.4.8. Construct the Seifert diagrams of the knots T_3 and $P_{-3,2,3}$, pictured in Figure 3.4.12.

Figure 3.4.12: The knots T_3 and $P_{-3,2,3}$.

 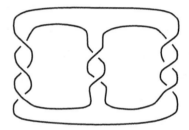

Next, we look at various Seifert diagrams of knots that are not presented in closed braid form and look for ways to manipulate them into a closed braid form. There is one important caveat. We must take care when manipulating a Seifert diagram to ensure that the resulting diagram represents an equivalent knot. Our goal is for the Seifert circles within the final diagram to be concentric and coherently oriented.

Exercise 3.4.9. Figure 3.4.13 represents the Seifert circle representation of a knot diagram. This diagram is not a closed braid because the circle labeled **C** fails to be concentric. Show that the diagram can be altered slightly to represent a closed braid. Prove that your alteration preserves knot equivalence by drawing the knot diagrams before and after the alteration and describing the Reidemeister moves that prove these two diagrams are equivalent.

Figure 3.4.13: This diagram is *almost* in closed braid form.

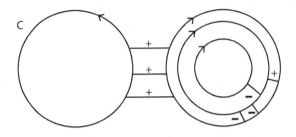

Exercise 3.4.10. Generalize your proof above to the following, more general scenario. Suppose a Seifert diagram contains two closed braid subsections that are oppositely oriented and connected by a collection of crossings between the outermost circles of each subsection (for example, see Figure 3.4.14). Prove that such a diagram is equivalent to that of a closed braid. (Hint: Recall that Reidemeister moves are not defined on Seifert diagrams. Thus, an argument proving equivalence must use the knot diagrams, not the Seifert diagrams.) Your proof should be a general argument that does not rely on Figure 3.4.14 specifically.)

Figure 3.4.14: This diagram is not in closed braid form. Can it be altered to produce an equivalent diagram that is in closed braid form? Investigate!

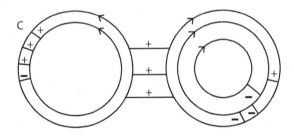

There are other types of Seifert diagrams that we have yet to consider, for example, the Seifert diagram of the knot T_3 from Exercise 3.4.8. The Seifert diagram for this knot has four circles and none are concentric. What do we do in this case? Let's simplify this problem and consider a pair of Seifert circles. Once we gain some intuition, we can try to generalize our results so that they apply to the Seifert diagram T_3.

There are two cases to consider when looking at a pair of non-nested Seifert circles: either the two circles have the same orientation or they have opposite orientation (one clockwise and one counterclockwise). Our previous argument in Exercise 3.4.10 can be used to address the latter case, so let's focus on the case where the two circles have the same orientation, as in Figure 3.4.15.

Figure 3.4.15: Two Seifert circles, both with clockwise orientation.

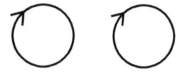

To put the circles in Figure 3.4.15 into a closed braid form, the obvious thing to do is to pick up one circle and place it inside the other. Although this idea will work in this simple case, it is not generalizable to the case where the two Seifert circles are connected to other circles via signed-line segments. So instead, let's use an R2 move on a portion of each circle. Since the circular arcs in a Seifert diagram represent the strands of a link, we can perform an R2 move in a region adjacent to the circles where no signed arcs are located. When an R2 move is performed, there will be two new crossings introduced that must then be smoothed using the Seifert diagram technique to produce a Seifert diagram.

Exercise 3.4.11. Determine the Seifert diagram associated to Figure 3.4.15 after an R2 move is applied.

Exercise 3.4.12. Transform the Seifert diagram of the twist knot T_3 into a closed braid form. (Hint: Pair up similarly oriented but nonconcentric Seifert circles and use the R2 move strategy from Exercise 3.4.11. You may also need the techniques from Exercise 3.4.10 to create your final braid diagram.)

Exercise 3.4.13. Prove that every twist knot can be transformed into a closed braid. (You are encouraged to think about this proof as *precisely describing* an algorithm that will transform the Seifert diagram of a twist knot into closed braid form. Throughout the algorithm, you should explain not only *how* to apply the algorithm, but *why* you know that you will be able to apply the algorithm and *why* you know your algorithm will terminate.)

As we work toward proving that every knot can be represented as a closed braid, we introduce new terminology for pairs of Seifert circles and a measurement of how far away a Seifert diagram is from closed braid form. A pair of circles within a Seifert diagram is called an **incoherently oriented pair** if the circles are nested and have opposite orientation, or if they are not nested and have the same orientation. In Figure 3.4.16, we see two examples of incoherently oriented pairs. Such a pair of circles would *not* be seen in a Seifert diagram that is in closed braid form. On the other hand, the pairs of circles in Figure 3.4.17 are said to be **coherently oriented pairs**. These pairs are either nested circles with the same orientation, or non-nested circles with opposite orientation. Notice that any two circles in a closed braid diagram are a coherently oriented pair.

Figure 3.4.16: Two pairs of incoherently oriented circles.

Figure 3.4.17: Two pairs of coherently oriented circles.

The following exercise shows that some diagrams of oriented circles connected by signed segments cannot be attained as Seifert diagrams of oriented knots or links.

Exercise 3.4.14. Prove that a pair of incoherently oriented circles within a Seifert diagram cannot be connected by a signed arc. (Note that there are a total of four distinct pairs of incoherently oriented circles, two of which are in Figure 3.4.16.)

Exercise 3.4.15. Use Exercise 3.4.14 to show that the diagram in Figure 3.4.18 cannot be part of a Seifert diagram, regardless of how orientations are assigned to the circles.

Figure 3.4.18: This diagram cannot be part of a Seifert diagram.

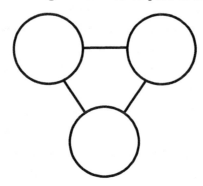

Next, we define a measurement of how far away a Seifert diagram is from closed braid form.

Definition 3.4.16. The **height of a Seifert diagram** D, denoted $h(D)$, is the number of distinct pairs of incoherently oriented Seifert circles in D.

Notice that $h(D) = 0$ if and only if D is (1) in closed braid form or (2) in the generalized form of Figure 3.4.14. As we saw in Exercise 3.4.10, the diagram in Figure 3.4.14 is just a short manipulation away from being in closed braid form.

Exercise 3.4.17. Show that the Seifert diagram for the knot T_4 in Figure 3.4.11 has height 3.

Exercise 3.4.18. Draw a Seifert diagram of a knot that has height 1 and a Seifert diagram of a knot that has height 2. How many such diagrams of height 1 or 2 can you find? (Make sure your diagrams are *legal* Seifert diagrams by applying Exercise 3.4.14.)

Exercise 3.4.19. Determine the height of the Seifert diagram in Figure 3.4.19.

We know that if a Seifert diagram contains a pair of incoherently oriented circles, then it cannot be in closed braid form. The R2 move described in Exercise 3.4.11 shows how a pair of non-nested, incoherently oriented circles can be transformed into a pair of coherently oriented circles. The next exercise shows that whenever such an R2 move can be performed, doing so will reduce the height of the Seifert diagram.

Figure 3.4.19: A Seifert diagram.

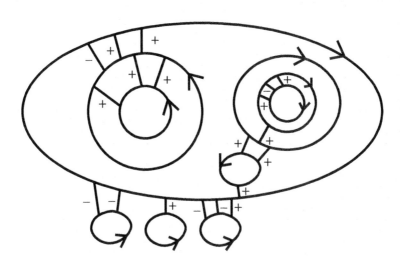

Exercise 3.4.20. Suppose C_1 and C_2 are an incoherently oriented pair of Seifert circles within a Seifert diagram D that are adjacent to a common region of the Seifert diagram. Prove that performing an R2 move on C_1 and C_2 within the common region results in a Seifert diagram D' with $h(D') < h(D)$.

We are very close to proving that every knot can be represented as a closed braid! There is only one more piece in our puzzle. We need to show that in every Seifert diagram with non-zero height, there exists a region adjacent to a pair of incoherently oriented Seifert circles in which we can apply an R2 move.

Definition 3.4.21. A **defect region** of a Seifert diagram is a region of a Seifert diagram that is adjacent to two incoherently oriented circles.

An example of a defect region is shown in Figure 3.4.20. The red dotted curves in Figure 3.4.21 show two possible places where an R2 move can be applied to an incoherently oriented pair of Seifert circles.

At last, we prove the final step, that if a Seifert diagram contains a pair of incoherently oriented circles, then it will also contain a defect region.

Figure 3.4.20: The shaded region is a defect region of the Seifert diagram. It happens to be adjacent to two pairs of incoherently oriented circles.

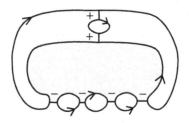

Figure 3.4.21: The height-reducing R2 move can be applied along either of the red dotted curves shown in these figures.

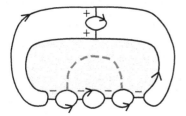 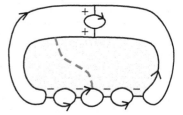

Exercise 3.4.22. Let K be an arbitrary knot with Seifert diagram D. If $h(D) > 0$, then D contains a defect region.

We now have all the necessary ingredients to prove what is known as Alexander's Theorem. This result was originally proved by James W. Alexander II in 1923 [3].

Theorem 3.4.23. [**Alexander's Theorem**] *Every knot can be represented as a closed braid.*

Exercise 3.4.24. Prove Alexander's Theorem.

Chapter 4

Knot Notation

In Section 1.7, you created your own knot notation that could be used to encode the information contained in an alternating knot diagram. Perhaps your knot notation was even flexible enough to encode the information in *any* knot diagram. If so, you may have discovered one of the many notations commonly used to encode knots by knot theorists today! In this chapter, we focus on several common ways to encode knotting information. Some of these codes involve strings of symbols, and some will require us to draw new types of diagrams.

4.1 DT Notation

One of the most common notations in use today is called **DT notation**, or **Dowker-Thistlethwaite notation**, named after knot theorists Clifford Hugh Dowker and Morwen Thistlethwaite. Here's how it works.

Starting at a given point on the knot diagram (which we will call the **base point**), travel in the direction of the orientation, labeling the first crossing you encounter with a 1, the second crossing you encounter with a 2, and so forth. Continue this process until every crossing has two labels and you have returned to the base point. Now, travel through your knot diagram a second time. For each time you pass *over* a crossing where you had placed an even number, make this even label negative. Even labels that were assigned as you passed *under* a crossing should remain positive.

For example, see Figure 4.1.1. When we labeled the second crossing, we were passing over the crossing, so our label should be -2. On the other hand, when we assigned a label to the fourth crossing, we were passing under the crossing, so the label should remain a positive number.

Now that we have our diagram labeled, we create an n-tuple of ordered pairs where the first element of each ordered pair is an odd number, and the second element is even. The number of elements n in this tuple is equal to the number of crossings in our diagram. Here's how we create this ordered list. The first element of the n-tuple should be $(1, m)$ for whichever integer m is paired with 1 in the diagram. For the example in Figure 4.1.1, for instance, m would equal -6. The second element of the n-tuple should be of the form $(3, q)$ where q is the integer paired with 3, etc.

Figure 4.1.1: A knot diagram with a DT-labeling.

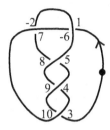

The 5-tuple we would create from Example 4.1.1 would be

$$((1, -6), (3, 10), (5, 8), (7, -2), (9, 4)).$$

Of course, when we form such a list, the odd numbers become entirely superfluous. For any n-tuple we create in this way, the first elements of the ordered pairs will always be the first n odd numbers, in order. So the final DT code we derive to represent the knot diagram is simply the ordered n-tuple of *even* numbers in the list, e.g., $(-6, 10, 8, -2, 4)$.

Exercise 4.1.1. Explain why labeling a knot diagram using the DT algorithm always gives you one odd and one even number at each crossing.

Exercise 4.1.2. Reconstruct the knot diagram that has DT code $(-6, -8, 12, -2, 14, 16, 4, 10)$. Is the knot diagram alternating?

Exercise 4.1.3. Find the DT code for knots (b) and (c) from Table 1.1 as follows. For both knots, select a base point and an orientation so that the first crossing you encounter, you pass over the crossing. What conclusions can you make from this exercise?

Exercise 4.1.4. Can you identify whether or not a knot diagram is alternating by looking at a DT code for the diagram? If so, explain how. If not, why not?

Exercise 4.1.5. Draw your favorite knot diagram and find the diagram's DT code. Have a friend draw their favorite knot diagram and do the same. Now, each of you should make a copy of your diagram, without labeling the crossings with DT code numbering and without specifying the starting point. You may, however, mark the orientation of your knot diagram. Finally, swap diagrams without giving away your notation. Find the DT codes for each other's diagrams. Did you come up with the same codes? Why or why not?

Exercise 4.1.6. Explore the effect on the DT code of moving your base point along the knot diagram past a single crossing.

For the next exercises, it will be helpful to use the following terminology.

Definition 4.1.7. A knot together with a base point is referred to as a **based knot**. Similarly, a knot diagram with a specified base point is referred to as a **based-knot diagram**.

Exercise 4.1.8. How does the DT code of an oriented-based-knot diagram D compare to the DT code of the same based-knot diagram with the opposite orientation? Formulate and prove a conjecture about this relationship.

Exercise 4.1.9. How does the DT code of an oriented-based-knot diagram D compare to the DT code of the mirror image of D? Formulate and prove a conjecture about this relationship.

We have seen that Reidemeister moves don't affect the knot type, but they do affect the knot diagram. Since the DT code is dependent on the specific diagram of the knot we are encoding, let's think about how the DT code changes as we perform Reidemeister moves.

Figure 4.1.2: Equivalent knot diagrams. From left to right, each diagram differs from the preceeding diagram by exactly one Reidemeister move.

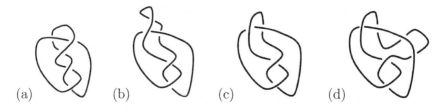

(a)   (b)   (c)   (d)

Exercise 4.1.10. Determine which Reidemeister moves have been performed to obtain each diagram in the sequence in Figure 4.1.2 from the first diagram. Then find the DT codes corresponding to these knot diagrams for a consistent choice of base point.

Exercise 4.1.11. Determine the effect on the DT code of performing an R1 move on a knot diagram.

Exercise 4.1.12. Determine the effect on the DT code of performing an R2 move on a knot diagram.

Exercise 4.1.13. Determine the effect on the DT code of performing an R3 move on a knot diagram.

Before moving on to the next way of encoding knotting information, discuss with a friend the pros and cons of using DT notation to encode knotting information.

4.2 Gauss Codes & Gauss Diagrams

So far, we have been considering knots by looking at their knot diagrams in the plane. There is another, quite useful *visual* way of representing a knot called a **Gauss diagram**. A Gauss diagram is made from a circle together with several decorated chords. More specifically, each chord is decorated with an arrowhead and a sign. Here is an example.

Figure 4.2.1: A knot diagram and its corresponding Gauss diagram.

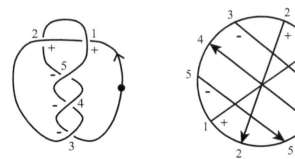

So what exactly is the correspondence between these two types of diagrams? Let's figure it out by focusing on our example. Begin by taking a look at the knot diagram in Figure 4.2.1. Find the base point on the diagram, and notice that each of the crossings is labeled by a single

number. Now, travel from the base point around the knot diagram in the
direction of the orientation. As you travel, record the order in which you
encounter the crossings. You might record this order in a list, such as:

$$1\ 2\ 3\ 4\ 5\ 1\ 2\ 5\ 4\ 3.$$

You might even record a bit more information by noting when you pass
over (O) and when you pass under (U) each crossing. For instance, you
could write the following:

$$U1\ O2\ O3\ U4\ O5\ O1\ U2\ U5\ O4\ U3.$$

This sequence is called a **Gauss code** and can already be used to encode a
great deal of information about the knot.

Exercise 4.2.1. Find a Gauss code for the the pretzel knot in Figure 4.2.2.

Figure 4.2.2: A pretzel-knot diagram.

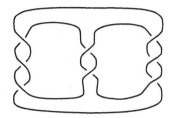

Let's return to our original example from Figure 4.2.1. To create a Gauss
diagram from this code, separately draw a circle with a base point on it.
Travel around the circle counterclockwise from the base point, listing the
numbers 1, 2, 3, 4, 5, 1, 2, 5, 4, 3 at regular intervals around the circle as
you go. You should have two 1s, two 2s, etc. Connect each of these pairs of
numbers by a chord. What you have now is a chord diagram, the first step
to creating your Gauss diagram!

You already recorded in your Gauss code when you passed over and when
you passed under each crossing. To record this same information in the
Gauss diagram, decorate the chords in your chord diagram with
arrowheads, making the arrow point toward the under-strand of the
crossing. For instance, you first pass *under* crossing 1 when you travel

around the knot from the base point. Thus, your arrow should be pointing *toward* your first encounter with chord 1 in the Gauss diagram. Add arrowheads to all chords in your chord diagram and then compare it with Figure 4.2.1 to see if you're on the right track.

Finally, determine the sign of each crossing, as in Definition 3.4.6, and decorate each arrow of your Gauss diagram with its sign. *Voila!* You are done. Your Gauss diagram should look the same as the one in Figure 4.2.1.

Exercise 4.2.2. Create a Gauss diagram for the pretzel knot in Figure 4.2.2.

Exercise 4.2.3. Create Gauss diagrams for all of the *knot* diagrams in Figure 1.1. (You may ignore any links with more than one component!)

Exercise 4.2.4. Suppose you are given a Gauss diagram that is associated to a knot diagram, but you aren't sure what the knot diagram looks like. Can you tell by looking at the Gauss diagram whether or not the corresponding knot diagram is alternating?

You may ask yourself a natural question at this point. Do different knot diagrams of the same knot correspond to different Gauss diagrams? To answer this question, let us consider the effect on the corresponding Gauss diagrams of performing a Reidemeister move on a knot diagram.

Exercise 4.2.5. Find the Gauss diagrams corresponding to the four equivalent knot diagrams in Figure 4.1.2. What can you conclude about the effect of Reidemeister moves in a Gauss diagram?

Using the previous exercise for intuition, let's think more systematically about the effects different R-moves have on Gauss diagrams of knots.

Exercise 4.2.6. Determine all possible effects on the corresponding Gauss diagram of performing an R1 move on a knot diagram.

Exercise 4.2.7. Show that if a Gauss diagram contains only nonintersecting arrows, as in Figure 4.2.3, then it must be a Gauss diagram of the unknot.

Exercise 4.2.8. In Figure 4.2.4, we see a Gauss diagram schema that illustrates one possible effect of an R2 move on a Gauss diagram. (The

Figure 4.2.3: A Gauss diagram containing only nonintersecting arrows.

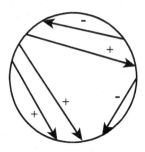

dotted portions of the circle in the Gauss diagram indicate that *any*
additional chord endpoints may appear in this portion of the diagram. The
solid portion of the circle in the Gauss diagram indicates that *no
additional* chord endpoints appear in this portion of the diagram.) Are
there other Gauss diagrammatic R2 moves? Determine all possible effects
on the corresponding Gauss diagram of performing an R2 move on a knot
diagram. Share your findings by providing a complete list of one or more
Gauss diagrammatic schemas, including the one shown in Figure 4.2.4.

Figure 4.2.4: A Gauss diagrammatic Reidemeister 2 move.

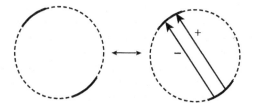

Exercise 4.2.9. Determine all possible effects on the corresponding Gauss
diagram of performing an R3 move on a knot diagram. Record your
findings by providing Gauss diagrammatic schemas, as in Figure 4.2.4, for
each R3 move.

Exercise 4.2.10. Compare the results you obtained in the previous
exercises to the Gauss diagrammatic R-moves illustrated in Figure 4.2.5.
Did you discover any possibilities that are missing in Figure 4.2.5?

As we learned in Section 2.3, *all* oriented R-moves can be derived from the
generating set of moves in Figure 2.3.6. So while you may have found

Figure 4.2.5: Gauss diagrammatic Reidemeister moves. The n is a variable that may represent $+$ or $-$.

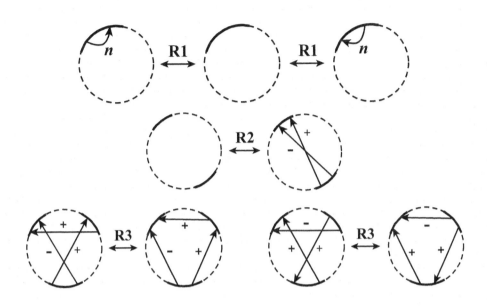

several related valid Gauss diagrammatic R-moves, the moves in Figure 4.2.5 will suffice to generate all possible equivalences of Gauss diagrams.

Figure 4.2.6: A Gauss diagrammatic R3 move that does not appear in Figure 4.2.5.

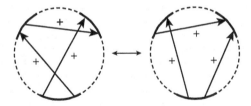

Exercise 4.2.11. Notice that the Gauss diagrammatic R2 move shown in Figure 4.2.4 and the R3 move shown in Figure 4.2.6 are not pictured in Figure 4.2.5. Prove that the two moves shown in Figures 4.2.4 and 4.2.6 can be obtained via a sequence of Gauss diagrammatic R-moves that are all in Figure 4.2.5. Your proofs must respect the dotted regions of the

circle in the Gauss diagram, meaning that your argument cannot assume
no other chords are within the dotted regions.

Exercise 4.2.12. You just learned how to make Gauss codes and Gauss
diagrams for knots, but what about links with more than one component?
Invent an extension of Gauss notation for two-component links.

4.3   Rational Tangles and Knots, and Conway Notation

Another type of knot notation, called **Conway notation**, is rather
different than the notations we've discussed so far. Conway notation can
be more difficult to determine for an arbitrary diagram, but it is incredibly
useful for defining and studying an interesting family of knots called
rational knots. We will narrow our focus here and use Conway notation
to describe only rational knots. Readers that are interested in exploring
Conway notation for nonrational knots can find more information here [9].

The building blocks of rational knots, called *tangles*, are special types of
subdiagrams of a knot or link diagram.

Definition 4.3.1. A **tangle diagram** is a diagram contained in a disk
that consists of two arcs whose four endpoints are fixed along the
boundary of the disk. The disk is denoted by a dashed circle surrounding
the subdiagram. Sometimes the dashed disk boundary is omitted, but
implied. When drawing a tangle, we situate the four fixed endpoints
roughly on the NW, NE, SE, and SW compass points.

Figure 4.3.1: Examples of tangle diagrams.

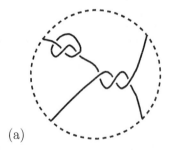
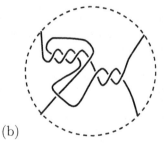

(a)   (b)

We can view a tangle diagram as a subdiagram of a knot or link by imagining that the two arcs in the tangle extend into a larger knot or link diagram outside of the disk.

Definition 4.3.2. Two tangle diagrams are **equivalent** provided that one can be continuously deformed into the other via planar isotopies and Reidemeister moves performed within the dashed circle *while keeping the arc endpoints fixed*. The symbol $\sim$ denotes tangle equivalence.

Exercise 4.3.3. To practice working with tangle equivalence, find two pairs of equivalent tangles in Figure 4.3.2. Remember, when manipulating a tangle, the four endpoints must remain fixed.

Figure 4.3.2: Which pairs of tangles are equivalent?

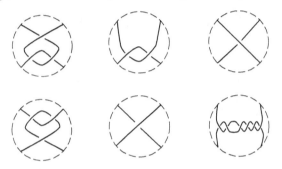

Conway's notation for rational knots elucidates the rich algebraic and geometric structures that stem from combining and manipulating tangles. Tangles can be summed, multiplied, rotated, mirrored, inverted, and flipped.

Definition 4.3.4. Let T, T_1, T_2, and Q be tangles. The **tangle sum** of T_1 and T_2, denoted $T_1 + T_2$ and depicted in Figure 4.3.3, is the result of horizontally arranging T_1 and T_2 and connecting their adjacent arc endpoints. The **tangle product** of T_1 and T_2, denoted $T_1 * T_2$ and depicted in Figure 4.3.4, is the result of vertically stacking T_1 and T_2 and connecting adjacent arc endpoints. The **rotation of a tangle** T, denoted T^r, rotates T clockwise by 90 degrees. The **mirror of a tangle** T, denoted $-T$, is the tangle resulting from switching all crossings in T. The **inverse of a tangle** T, denoted T^i, is defined to be $-T^r$, the mirror of the clockwise 90-degree rotation of T. The last two operations on a tangle Q are the **vertical** and **horizontal flips** of Q over a vertical or horizontal line, denoted by Q^{Vflip} and Q^{Hflip}, respectively, and depicted in Figure 4.3.6.

Figure 4.3.3: The sum of two tangles, $T_1 + T_2$.

Figure 4.3.4: The product of two tangles, $T_1 * T_2$.

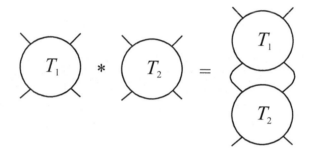

Figure 4.3.5: The rotation and inverse of a tangle.

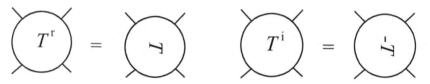

The tangle building blocks that are used to create all rational tangles are the **integer** and **reciprocal** tangles, examples of which are shown in Figures 4.3.7 and 4.3.8.

Definition 4.3.5. We define a **rational tangle** as a tangle that can be constructed inductively, as follows. All integer and reciprocal tangles are rational tangles. Given a rational tangle T_k, the tangle T_{k+1} that is obtained from T_k by either (1) *adding an integer tangle* or (2) *multiplying by a reciprocal tangle* is also a rational tangle. A diagram of a rational tangle, T, that depicts the finite sequence in the inductive construction of T is called a **twist diagram** of the rational tangle T.

Figure 4.3.6: The vertical and horizontal flip of a tangle.

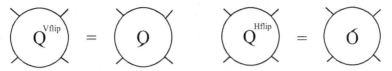

Figure 4.3.7: Examples of integer tangles.

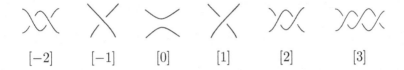

$[-2]$ $\qquad$ $[-1]$ $\qquad$ $[0]$ $\qquad$ $[1]$ $\qquad$ $[2]$ $\qquad$ $[3]$

Figure 4.3.8: Examples of reciprocal tangles.

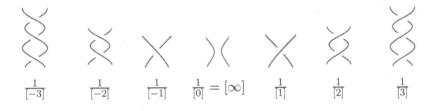

$\frac{1}{[-3]}$ $\quad$ $\frac{1}{[-2]}$ $\quad$ $\frac{1}{[-1]}$ $\quad$ $\frac{1}{[0]} = [\infty]$ $\quad$ $\frac{1}{[1]}$ $\quad$ $\frac{1}{[2]}$ $\quad$ $\frac{1}{[3]}$

Figure 4.3.9: Some twist diagrams of rational tangles.

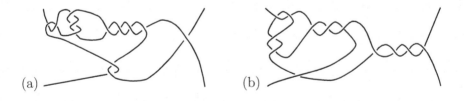

(a) $\qquad\qquad\qquad\qquad$ (b)

When first looking at a twist diagram, it can be difficult to determine the sequence of integer and reciprocal tangles used to make the rational tangle. However, if you are given a twist diagram, you will be able to identify the *last* summand or factor because two adjacent arc endpoints (from that last summand or factor) can be untwisted to remove it from the tangle. In other words, from T_{k+1}, we can always find two adjacent arc endpoints in the tangle that can be untwisted (some number of times) to uncover T_k.

The second to last summand or factor can be identified similarly, and so on, until the starting point of the rational tangle is apparent. For example, we can find the construction sequence for rational tangle (a) in Figure 4.3.9 by observing it has a *last* summand of $[-1]$ on the right of the diagram. Untwisting this rightmost $[-1]$ summand, we see the second to last term was multiplication below by the reciprocal tangle $\frac{1}{[-2]}$. Untwisting the factor $\frac{1}{[-2]}$, we see a left summand of $[2]$ and a right summand of $[-4]$. Untwisting both integer tangles $[2]$ and $[-4]$, we see that the rational tangle (a) was constructed starting with the reciprocal tangle $\frac{1}{[3]}$. Finally, we arrive at the following notation for the rational tangle (a) in Figure 4.3.9:

$$\left(\left([2]+\frac{1}{[3]}+[-4]\right)*\frac{1}{[-2]}\right)+[-1].$$

Exercise 4.3.6. Write the rational tangle (b) in Figure 4.3.9 as a sum and product of integer and reciprocal tangles by using the untwisting strategy described above. Be careful to use parentheses appropriately in your final notation!

In the next exercise, we will show that the notation used for describing a twist diagram is not unique. For example, the rational tangle (a) in Figure 4.3.9 can also be described as:

$$\left(\left([2]+\frac{1}{[3]}+[-1]+[-3]\right)*\frac{1}{[-2]}\right)+[-1]$$

or, by an equivalence, as

$$\left(\left([2]+\frac{1}{[3]}+[-1]+[-3]\right)*\frac{1}{[-2]}\right)+[-6]+[5].$$

Exercise 4.3.7. Prove the following three statements.

1. For all integers n and m, $[n]+[m]\sim[n+m]$.

2. For all integers n and m, $\frac{1}{[n]}*\frac{1}{[m]}\sim\frac{1}{[n+m]}$.

3. For any nonzero integer n, $[n]^i=\frac{1}{[n]}$ and $\left(\frac{1}{[n]}\right)^i=[n]$. Also, $[0]^i=[\infty]$ and $[\infty]^i=[0]$.

Now that we are familiar with the construction of rational tangles, a reasonable question to ask is *why* do we restrict to summing integer tangles and multiplying by reciprocal tangles? Can't we add two reciprocal tangles? Or multiply by an integer tangle? Well, in truth, both of these things can be done; however, the result is generally not a rational tangle. One of the key properties of a rational tangle is that, if we allow the endpoints of the tangle to move, it can be completely untwisted until it becomes the [0] or [∞] tangle. In Figure 4.3.10 there are some tangles that fail to have this property when we use the sum operation with reciprocal tangles or multiplication with integer tangles.

Exercise 4.3.8. Identify the tangle sums from Figure 4.3.10 that are *not* rational.

Figure 4.3.10: Four examples of tangle sums and products, not all of which are rational.

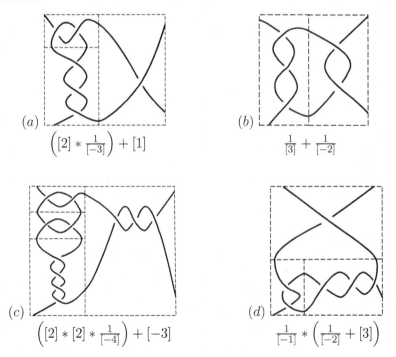

(a) $\left([2] * \frac{1}{[-3]} \right) + [1]$

(b) $\frac{1}{[3]} + \frac{1}{[-2]}$

(c) $\left([2] * [2] * \frac{1}{[-4]} \right) + [-3]$

(d) $\frac{1}{[-1]} * \left(\frac{1}{[-2]} + [3] \right)$

Exercise 4.3.9. Identify the tangle in Figure 4.3.1 that is *not* a rational tangle.

Exercise 4.3.10. For each tangle, determine whether or not it is rational.

(i) $\left([-5] + \left(\left(\frac{1}{[4]} * [5]\right) * \frac{1}{[-3]}\right)\right) * \frac{1}{[10]}$

(ii) $\left(\frac{1}{[6]} * \left([-5] + \frac{1}{[4]} + [5]\right) * \frac{1}{[-3]}\right) + [-2]$

With the next exercise and definition, we set the stage to uncover some surprising results about the vertical and horizontal flip of a rational tangle.

Exercise 4.3.11. Prove that integer and reciprocal tangles are equivalent to their flips. That is, prove the following four equivalences.

1. $([n])^{Vflip} \sim [n]$ 3. $\left(\frac{1}{[m]}\right)^{Vflip} \sim \frac{1}{[m]}$

2. $([n])^{Hflip} \sim [n]$ 4. $\left(\frac{1}{[m]}\right)^{Hflip} \sim \frac{1}{[m]}$

Definition 4.3.12. Given a tangle of the form $[\pm 1] + P$, the horizontal flip of P that untwists $[\pm 1]$ and yields the equivalent tangle $P^{Hflip} + [\pm 1]$ is called a **flype**. Similarly, for a tangle of the form $\frac{1}{[\pm 1]} * P$, the vertical flip of P that untwists $\frac{1}{[\pm 1]}$ and yields the equivalent tangle $P^{Vflip} * \frac{1}{[\pm 1]}$ is also called a **flype**. Two of these four flype equivalences are depicted in Figure 4.3.11.

Figure 4.3.11: The flype equivalences.

$[-1] + P \sim P^{Hflip} + [-1]$ $P * \frac{1}{[1]} \sim \frac{1}{[1]} * P^{Vflip}$

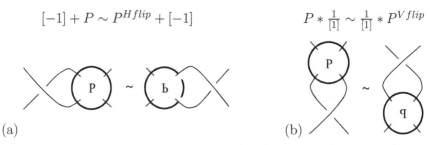

(a)   (b)

In Exercise 4.3.11, we saw that the flips of integer and reciprocal tangles are equivalent to themselves. Perhaps surprisingly, *all* rational tangles are equivalent to their flips! Also, another interesting result follows as a corollary of the flip equivalence, namely that tangle addition and multiplication are, in a sense, commutative.

Theorem 4.3.13. [**The Flip Theorem**] *For a rational tangle P,*

$$P \sim P^{Vflip} \quad and \quad P \sim P^{Hflip}.$$

Corollary 4.3.14. *For rational tangle P and an integer m,*

$$P + [m] \sim [m] + P \quad and \quad P * \frac{1}{[m]} \sim \frac{1}{[m]} * P.$$

The next exercise and the example in Figure 4.3.12 help us build intuition that will lead to the proof of The Flip Theorem and its corollary.

Exercise 4.3.15. Use the flype equivalences to prove that the following statements hold for all integers n and m.

(i) $[n] + \frac{1}{[m]} \sim \frac{1}{[m]} + [n]$, and

(ii) $[n] * \frac{1}{[m]} \sim \frac{1}{[m]} * [n]$.

Figure 4.3.12: An example showing $P^{Hflip} \sim P$, for a rational tangle, P.

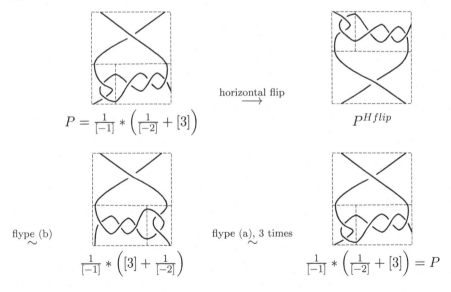

$$P = \frac{1}{[-1]} * \left(\frac{1}{[-2]} + [3] \right)$$

horizontal flip $\longrightarrow$

P^{Hflip}

flype (b) $\underset{\sim}{}$

$$\frac{1}{[-1]} * \left([3] + \frac{1}{[-2]} \right)$$

flype (a), 3 times $\underset{\sim}{}$

$$\frac{1}{[-1]} * \left(\frac{1}{[-2]} + [3] \right) = P$$

As seen in Figure 4.3.12, for the rational tangle $P = \frac{1}{[-1]} * \left(\frac{1}{[-2]} + [3] \right)$, its horizontal flip is given by $P^{Hflip} = \left(\frac{1}{[-2]} + [3] \right) * \frac{1}{[-1]}$. Next we use the flype equivalences of type (a) and (b) from Figure 4.3.11. First, we apply a flype of type (b) to $P^{H\,flip}$, moving the factor $\frac{1}{[-1]}$ from the bottom to the top of the figure. Then, three flypes of type (a) transform $[3] + \frac{1}{[-2]}$ to $\frac{1}{[-2]} + [3]$, and this completes the diagrammatic argument showing $P^{H\,flip} \sim P$.

Exercise 4.3.16. Using the techniques shown in Figure 4.3.12, write a sequence of diagram equivalences that show $Q \sim Q^{V\,flip}$ for $Q = \left(\frac{1}{[-1]} * \left([3] + \frac{1}{[2]} \right) \right) + [-2]$.

Next we will prove The Flip Theorem using the intuition we have gained from Figure 4.3.12 and Exercise 4.3.16.

Exercise 4.3.17. Prove The Flip Theorem. (Hint: Use induction on the number of crossings in P. For the inductive step there will be four cases to consider: $P = Q + [\pm 1]$, $P = [\pm 1] + Q$, $P = Q * \frac{1}{[\pm 1]}$, and $P = \frac{1}{[\pm 1]} * Q$. Notice that Q is a rational tangle with one fewer crossing than P.)

Exercise 4.3.18. Use The Flip Theorem to prove Corollary 4.3.14.

The next equivalence is simple, but eye opening! Recall from part 3 of Exercise 4.3.7, we recognized that a reciprocal tangle $\frac{1}{[a]}$ is the inverse of the integer tangle $[a]$, and vice versa. In Theorem 4.3.19, we recognize the product $P * \frac{1}{[a]}$ as the inverse of a sum. This equivalence is a key step in understanding the term *rational* is used in the name of rational tangles. From here onward, we use the notation $\frac{1}{P}$ instead of P^i.

Theorem 4.3.19 (The Product-to-Inverse Equivalence). *For a rational tangle P and an integer a,*

$$P * \frac{1}{[a]} \sim \frac{1}{[a] + \frac{1}{P}} \qquad\qquad \frac{1}{[a]} * P \sim \frac{1}{\frac{1}{P} + [a]}.$$

Exercise 4.3.20. Prove Theroem 4.3.19. (Hint: Draw the diagrams within the claimed equivalence and provide a Reidemeister sequence of diagrams that prove they are equivalent tangles.)

The Product-to-Inverse Equivalence will allow us to transform our notation for a rational tangle in a very important way; given the notation for a rational-tangle-twist form we can change it into notation that looks like a continued fraction. This change is accomplished by replacing each product expression with the equivalent inverse expression. As our notation for a rational tangle becomes more like a rational number, we will get closer to understanding a powerful association between rational tangles and rational numbers.

First, let's use an example to clarify the notational-transformation process of replacing product expressions with the equivalent inverse expressions. In this example, we use Corollary 4.3.14 and Theorem 4.3.19 repeatedly, and the results from Exercise 4.3.7.

Let's start with the notation for a particular rational tangle in twist form,

$$[5] + \left(\frac{1}{[4]} * \left([3] + (\frac{1}{[2]} * [6] * \frac{1}{[10]}) + [-8] \right) \right).$$

Using the Product-to-Inverse Equivalence, we can transform $\frac{1}{[4]} * T$ into the equivalent tangle $\frac{1}{[4]+\frac{1}{T}}$, where $T = [3] + (\frac{1}{[2]} * [6] * \frac{1}{[10]}) + [-8]$. Hence,

$$[5] + \left(\frac{1}{[4]} * ([3] + (\frac{1}{[2]} * [6] * \frac{1}{[10]}) + [-8])) \right)$$

$$\sim [5] + \cfrac{1}{[4] + \cfrac{1}{([3] + (\frac{1}{[2]} * [6] * \frac{1}{[10]}) + [-8])}}.$$

Next, from commutativity of adding $[-8]$ and the commutativity of multiplying by $\frac{1}{[2]}$, and from using that $[3] + [-8] \sim [-5]$ and $\frac{1}{[2]} * \frac{1}{[10]} \sim \frac{1}{[12]}$, we have the following two equivalences.

$$[5] + \cfrac{1}{[4] + \cfrac{1}{[3]+(\frac{1}{[2]}*[6]*\frac{1}{[10]})+[-8]}} \sim [5] + \cfrac{1}{[4] + \cfrac{1}{[-5] + (\frac{1}{[2]} * [6] * \frac{1}{[10]}))}}$$

$$\sim [5] + \cfrac{1}{[4] + \cfrac{1}{[-5] + ([6] * \frac{1}{[12]})}}$$

Using the Product-to-Inverse Equivalence again, this time on the product $[6] * \frac{1}{[12]}$, we have

$$[5] + \cfrac{1}{[4] + \cfrac{1}{[-5] + ([6] * \frac{1}{[12]})}} \quad \sim \quad [5] + \cfrac{1}{[4] + \cfrac{1}{[-5] + \cfrac{1}{[12] + \frac{1}{[6]}}}}.$$

Now, our rational tangle that was once in twist form looks rather like a continued fraction!

Exercise 4.3.21. (a) Use Corollary 4.3.14 and the Product-to-Inverse Equivalences to transform the rational tangle below into an expression that looks like a continued fraction.

$$\frac{1}{[5]} * \left([4] + \left(\frac{1}{[3]} * ([2] + \frac{1}{[6]} + [10]) * \frac{1}{[-8]} \right) \right)$$

(b) Compare your continued fraction from part (a) to the preceding example. The notation should suggest a relationship between the rational tangles

$$[5] + \left(\frac{1}{[4]} * \left([3] + \left(\frac{1}{[2]} * [6] * \frac{1}{[10]} \right) + [-8] \right) \right) \quad \text{and}$$

$$\frac{1}{[5]} * \left([4] + \left(\frac{1}{[3]} * \left([2] + \frac{1}{[6]} + [10] \right) * \frac{1}{[-8]} \right) \right).$$

Using this example for inspiration, conjecture and prove a more general tangle relationship.

Given a rational tangle in twist form, the process from Exercise 4.3.21 and the preceding example can be applied to write the tangle in what is called *continued fraction form*.

Definition 4.3.22. A **continued fraction form** of a rational tangle is an algebraic description of a rational tangle by a continued fraction of integer tangles as shown below

$$[a_n] + \cfrac{1}{[a_{n-1}] + \cdots + \cfrac{1}{[a_2] + \frac{1}{[a_1]}}},$$

where each a_i is a nonzero integer, except possibly a_n which could be zero.

Notice that the integer tangle $[a_n]$ is the last tangle in the construction of T. The twists in $[a_n]$ are the portion of the rational tangle where untwisting could start if one were to deconstruct the rational tangle. On the other hand, the tangle $[a_1]$ denotes the tangle to which all other integer and reciprocal tangles must be summed and multiplied; that is, $[a_1]$ is the first tangle in the construction of T. When drawing a rational tangle T using its continued fraction notation, it is helpful to notice whether $[a_1]$ is an integer or reciprocal tangle when T is given in twist form. Since the integer tangles in continued fraction notation alternate between reciprocal and integer tangles, we can simply use the parity of n (the number of integer tangles in the continued fraction form of T) to determine whether $[a_1]$ is an integer or reciprocal tangle in the twist notation for T.

We use these observations in the following exercise.

Exercise 4.3.23. Draw the rational tangle diagram from its continued fraction form.

$$\text{(a)} \quad [-4] + \cfrac{1}{[2] + \cfrac{1}{[-2] + \cfrac{1}{[5]}}} \qquad \text{(b)} \quad \cfrac{1}{[4] + \cfrac{1}{[3] + \cfrac{1}{[2] + \frac{1}{[5]}}}}$$

The continued fraction form of a rational tangle motivates the following association between continued fractions and rational tangles that was first seen to be a complete invariant for rational tangles by John H. Conway in 1970.

Definition 4.3.24. For a rational tangle T, with continued fraction form

$$[a_n] + \cfrac{1}{[a_{n-1}] + \cdots + \cfrac{1}{[a_2] + \frac{1}{[a_1]}}},$$

we defined the **fraction of** T, denoted $F(T)$, by

$$F(T) = F\left([a_n] + \cfrac{1}{[a_{n-1}] + \cdots + \cfrac{1}{[a_2] + \frac{1}{[a_1]}}}\right) := a_n + \cfrac{1}{a_{n-1} + \cdots + \cfrac{1}{a_2 + \frac{1}{a_1}}}.$$

Theorem 4.3.25. [**Conway's Theorem**] *Two rational tangles are equivalent if and only if their associated fractions are equal.*

Conway's Theorem implies that the fraction defined above is a complete invariant of rational tangles. That is, it tells us exactly when two tangles are and are not equivalent. As an example, consider the two continued fractions below.

Exercise 4.3.26. Show the two fractions below are equal.

$$3 + \cfrac{1}{-2 + \frac{1}{4}} \qquad\qquad 2 + \cfrac{1}{2 + \frac{1}{3}}$$

The two fractions in Exercise 4.3.26 are the fractions associated to the rational tangles,

$$[3] + \cfrac{1}{[-2] + \frac{1}{[4]}} \qquad\qquad [2] + \cfrac{1}{[2] + \frac{1}{[3]}}.$$

Conway's Theorem implies that, since the two numerical fractions are equal, the two rational tangles are equivalent. Figure 4.3.14 shows this through a sequence of equivalences (planar isotopy, Reidemeister moves, and flypes).

Exercise 4.3.27. For each equivalence in Figure 4.3.14, identify whether it is a planar isotopy, a flype, or sequence of several Reidemeister moves.

Exercise 4.3.28. Use Conway's Theorem to determine whether or not the two rational tangles shown in Figure 4.3.13 are equivalent.

Figure 4.3.13: Are these rational tangles equivalent?

Exercise 4.3.29. Use Conway's Theorem to determine whether or not the two rational tangles shown in Figure 4.3.15 are equivalent.

Figure 4.3.14: A sequence of tangle equivalences showing $\left([4] * \frac{1}{[-2]}\right) + [3] \sim \left([3] * \frac{1}{[2]}\right) + [2]$.

$$[3] + \frac{1}{[-2] + \frac{1}{[4]}} =$$

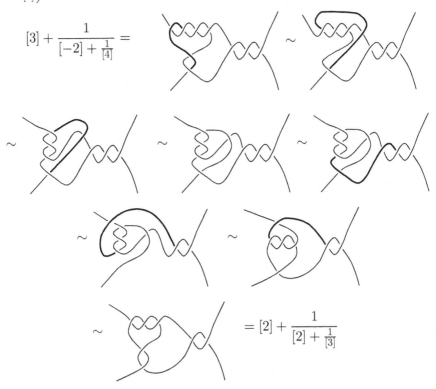

$$= [2] + \frac{1}{[2] + \frac{1}{[3]}}$$

Figure 4.3.15: Are these rational tangles equivalent?

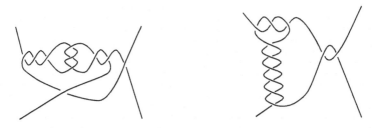

In what follows, our aim is to prove half of Conway's theorem, namely that if two rational tangles have equal fractions, then the rational tangles are equivalent. We direct the interested reader to several excellent sources that also prove the converse, that the fraction of a rational tangle is a well-defined function on equivalence classes of rational tangles [12,23].

Theorem 4.3.30. *If two rational tangles have equal fractions, then the tangles are equivalent.*

To prove that equal fractions imply equivalent tangles, we will use three ingredients: a unique representation of a continued fraction, an algebraic equality due to Lagrange, and a tangle equivalence that we will call "Lagrange's equivalence."

We begin with a fact about the unique representation of continued fractions. Every fraction, $\frac{p}{q}$, has a unique representation as

$$\frac{p}{q} = a_n + \cfrac{1}{a_{n-1} + \cfrac{1}{a_{n-2} + \ldots + \cfrac{1}{a_1}}}$$

such that all integers a_i, for $i < n$, are positive, (a_n could be negative or zero) and a_1 is greater than 1.

Such a continued fraction, written with all nonnegative integers except possibly a_n, will be called the **regular form continued fraction** of $\frac{p}{q}$. Notice that within a regular form continued fraction, the expression $\frac{1}{a_{n-1}+\ldots+\frac{1}{a_1}}$ is always positive and less than 1. This means that a_n must be the greatest integer less than or equal to $\frac{p}{q}$. Hence, given a fraction $\frac{p}{q}$, the value of a_n in its regular continued fraction form is unique. The following exercise should help make the existence and uniqueness of the remaining integers $a_{n-1}, a_{n-2}, \ldots, a_1$ seem plausible as well.

Exercise 4.3.31. For each of the three continued fractions below, show that it is equal to a regular form continued fraction. (Hint: First write the fraction as $\frac{p}{q}$, then determine the appropriate value for a_n. Next, determine $a_{n-1} > 0$, $a_{n-2} > 0$, and so on. Reflect on your algorithmic process to construct an argument for why the resulting regular continued fraction representation of $\frac{p}{q}$ is unique.)

(i) $4 + \cfrac{1}{-3 + \cfrac{1}{1 + \cfrac{1}{-4}}}$

(ii) $\dfrac{1}{1 + \dfrac{1}{-2 + \dfrac{1}{-3 + \dfrac{1}{4}}}}$

(iii) $-4 - \dfrac{1}{2 + \dfrac{1}{-2 + \dfrac{1}{-1 + \dfrac{1}{-5}}}}$

Next a useful formula, that is attributed to Lagrange, is given by

$$a - \frac{1}{b} = (a - 1) + \frac{1}{1 + \dfrac{1}{b - 1}}.$$

Exercise 4.3.32. Use algebra to prove that the formula of Lagrange, stated above, is a valid formula.

This formula, used first by Goldman and Kauffman in the proof of Conway's theorem [12], can be used to transform a continued fraction that is not in regular form into one that is in regular form. Observe that the negative term $-\frac{1}{b}$ on the left of the equation is transformed into an expression including the nonnegative term $+\frac{1}{b-1}$ on the right side. Thus, the right side of the equality has one fewer negative sign. A partial example that illustrates this transformation is shown below.

Consider the continued fraction (a) from Exercise 4.3.31, which is not in regular form. To partially transform this expression into its regular form, we first factor out the negative signs so that they become subtraction operations in the continued fraction expression. This is needed for the application of Lagrange's formula, as the left side of the equation involves a difference with the expression $\frac{1}{b}$ rather than the addition of $\frac{1}{-b}$.

$$4 + \dfrac{1}{-3 + \dfrac{1}{1 + \dfrac{1}{-4}}} = 4 - \dfrac{1}{3 - \dfrac{1}{1 - \dfrac{1}{4}}}$$

Notice that, after factoring out the *two* negative signs in front of the 3 and
4, we have a total of *three* subtractions. This means that we need only to
apply Lagrange's formula at most three times. We begin with the
subtraction at the bottom of the continued fraction.

$$4 - \cfrac{1}{3 - \cfrac{1}{1 - \cfrac{1}{4}}} = 4 - \cfrac{1}{3 - \cfrac{1}{(1-1) + \cfrac{1}{1 + \cfrac{1}{4-1}}}}$$

$$= 4 - \cfrac{1}{3 - \cfrac{1}{0 + \cfrac{1}{1 + \cfrac{1}{3}}}}$$

$$= 4 - \cfrac{1}{3 - \cfrac{1}{(1 + \cfrac{1}{3})}}$$

$$= 4 - \cfrac{1}{3 - \cfrac{1}{(1 + \cfrac{1}{3})}}$$

$$= 4 - \cfrac{1}{2 - \cfrac{1}{3}}$$

Exercise 4.3.33. Use Lagrange's formula twice more on the previous
example to find the regular form of this continued fraction. Note that your
final expression should be the same continued fraction as was found in
Exercise 4.3.31.

Exercise 4.3.34. Use the Lagrange formula to transform the continued
fractions (b) and (c) from Exercise 4.3.31 into regular form. (Hint: Factor
all negative signs out of fraction denominators to become subtraction signs.
Then apply Lagrange's formula multiple times from the bottom upward.)

Last, we state and then prove a tangle equivalence inspired by Lagrange's
equality; not only does Lagrange's formula hold for continued fractions,
but it also holds for rational tangles!

Theorem 4.3.35 (Lagrange's Equivalence for Rational Tangles). *For an
integer tangle* $[P]$ *and a rational tangle* $[Q]$,

$$[P] - \frac{1}{[Q]} \sim [P - 1] + \cfrac{1}{[1] + \cfrac{1}{[Q] + [-1]}}.$$

Exercise 4.3.36. Prove Lagrange's Equivalence for Rational Tangles. (Hint: Draw the diagram for $[P - 1] + \cfrac{1}{[1] + \frac{1}{[Q]+[-1]}}$ by first drawing $[Q] + [-1]$, then invert it, and so on. Next, draw the diagram for $[P] - \frac{1}{[Q]}$. Find a short sequence of equivalences (using planar isotopies, flypes, flips, or a sequence of Reidemeister moves) that transform the more complicated diagram into the simpler. State the name of each step of your equivalence as a planar isotopy, flype, flip, or sequence of Reidemeister moves. For inspiration, you may wish to look at the equivalences in Figure 4.3.14.)

The last step in the proof of Theorem 4.3.30 is to put these ideas together. We start with the assumption that T and R are two rational tangles whose associated fractions both equal $\frac{p}{q}$. Denote the continued fraction form of T and R by

$$T = [b_k] + \cfrac{1}{[b_{k-1}] + \cfrac{1}{[b_{k-2}]+...+\frac{1}{[b_1]}}} \qquad R = [c_n] + \cfrac{1}{[c_{n-1}] + \cfrac{1}{[c_{n-2}]+...+\frac{1}{[c_1]}}},$$

and denote the regular form of $\frac{p}{q}$ as $a_j + \cfrac{1}{a_{j-1} + \cfrac{1}{a_{j-2} + \cdots + \frac{1}{a_1}}}$.

Figure 4.3.16: The hypothesis of Theorem 4.3.30. Suppose T and R are rational tangles such that $F(T) = \frac{p}{q} = F(R)$.

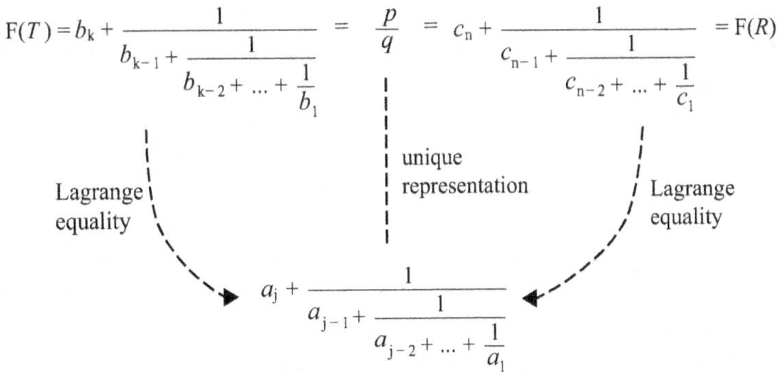

$$F(T) = b_k + \cfrac{1}{b_{k-1} + \cfrac{1}{b_{k-2} + ... + \frac{1}{b_1}}} = \frac{p}{q} = c_n + \cfrac{1}{c_{n-1} + \cfrac{1}{c_{n-2} + ... + \frac{1}{c_1}}} = F(R)$$

Lagrange equality · unique representation · Lagrange equality

$$a_j + \cfrac{1}{a_{j-1} + \cfrac{1}{a_{j-2} + ... + \frac{1}{a_1}}}$$

Exercise 4.3.37. Write a formal proof of Theorem 4.3.30. (Hint: Referring to Figure 4.3.16, what happens when Lagrange's equality is repeatedly applied to the continued fraction $F(T)$ with the goal of transforming it into a regular form continued fraction? What happens when Lagrange's equality is repeatedly applied to the continued fraction $F(R)$ with the goal of transforming it into a regular form continued fraction? How does this relate to $\frac{p}{q}$? To prove $T \sim R$, it suffices to prove $T \sim Q$ and $Q \sim R$ for some rational tangle Q. Figure 4.3.17 should inspire both the selection of Q and the means of proving that both T and R are equivalent to Q.)

Figure 4.3.17: The desired conclusion of Theorem 4.3.30 is to show that $T \sim R$.

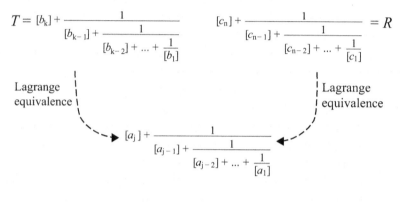

Now we will make some connections back to the world of knots and links by defining rational knots and links. There are two natural ways to connect the strand-ends of a tangle to create a knot or link. They are called the **numerator** and **denominator closures** of T, denoted by $N(T)$ and $D(T)$, respectively, and depicted in Figure 4.3.18.

Figure 4.3.18: The numerator and denominator closures of a tangle, T.

$$N(T) = \quad \boxed{T} \qquad\qquad D(T) = \quad \boxed{T}$$

Definition 4.3.38. A knot or link, L, is called **rational** provided that there exists a rational tangle T such that either $N(T)$ or $D(T)$ is equivalent to L.

Exercise 4.3.39. Prove or disprove each of the following statements.

1. For a tangle, T, $N(T) = D(T^r)$.

2. For a tangle, T, $D(T) = N(T^r)$.

3. For a rational tangle T, $N(T * \frac{1}{[a]}) = N(T)$.

4. For a rational tangle T, $D(T + [a]) = D(T)$.

Parts 3 and 4 of Exercise 4.3.39 imply that if K is a rational link such that $K = N(T)$ for a rational tangle T, then we can assume that the last tangle algebraically contributing to T is an integer tangle. Similarly, if K is a rational link such that $K = D(T)$ for some rational tangle T, then we may assume that the last tangle algebraically contributing to T is a reciprocal tangle.

As we mentioned at the beginning of this section, Conway notation is a useful notation for encoding rational links and knots. In fact, Conway used his notation (for rational and nonrational knots) to create a list (by hand) of all prime knots with 11 or fewer crossings.

Definition 4.3.40. Suppose the knot K is equivalent to the numerator closure of a rational tangle, T, with continued fraction form

$$T = [a_n] + \cfrac{1}{[a_{n-1}] + \cdots + \cfrac{1}{[a_2] + \frac{1}{[a_1]}}},$$

such that each a_i is a nonzero integer. The **Conway notation** for K is given by the string of integers $[[a_1 \ a_2 \ \ldots \ a_n]]$.

For example, the Conway notation $[[4 \ -2 \ 3 \ 1 \ 5]]$ encodes a knot whose final summand (before numerator closure) is the integer tangle $[5]$. Since the numbers listed alternate between integer and reciprocal tangles, the 3 and 4 values in the notation also represent integer tangles while the -2 and 1 represent reciprocal tangles. Hence the notation $[[4 \ -2 \ 3 \ 1 \ 5]]$ encodes the rational knot

$$K = N\left(\left(\left(\left([4] * \frac{1}{[-2]}\right) + [3]\right) * \frac{1}{[1]}\right) + [5]\right).$$

Similarly, to decode the notation [[3 6 −4 −3]], we observe that the final
summand of the rational tangle is the integer tangle [−3]. Hence, within
the notation [[3 6 −4 −3]], the 3 and −4 represent reciprocal tangles and 6
and −3 represent integer tangles. So [[3 6 −4 −3]] encodes the rational link

$$L = N \left(\left(\left(\frac{1}{[3]} + [6] \right) * \frac{1}{[-4]} \right) + [-3] \right).$$

Exercise 4.3.41. The Conway notation for all nontrivial prime knots with
seven crossings is given below. Draw the knots that correspond to each
sequence.

$$[[7]] \qquad [[5 \ 2]] \qquad [[4 \ 3]] \qquad [[3 \ 2 \ 2]] \qquad [[3 \ 1 \ 3]]$$

$$[[2 \ 2 \ 1 \ 2]] \qquad [[2 \ 1 \ 1 \ 1 \ 2]]$$

As a corollary of Conway's Theorem we have the following.

Corollary 4.3.42 (Conway). *Let L_1 and L_2 be rational links. Then L_1
and L_2 are equivalent if their continued fractions are equal.*

Exercise 4.3.43. Prove Corollary 4.3.42 using Theorem 4.3.25.

Note the difference between Conway's Theorem and its corollary. Not only
is every instance of "tangle" replaced by "link," but there is another
difference as well. Can you spot it? The corollary implies that there might
be equivalent rational links that have different associated continued
fractions. Perhaps this makes sense, because once we've closed a tangle, we
no longer have any endpoints we need to keep fixed. We have a bit more
freedom in how we might manipulate our link diagram. While there is
indeed more freedom, we state an incredibly useful result, proven by
Schubert, that completely characterizes rational link equivalence [36].

Theorem 4.3.44. [Schubert's Theorem] *Suppose L_1 is a rational link
with reduced fraction $\frac{p}{q}$ and L_2 is a rational link with reduced fraction $\frac{p'}{q'}$.
Then L_1 and L_2 are equivalent if and only if*

1. $p = p'$ and

2. either $q = q'$ (mod p) or $qq' = 1$ (mod p).

Exercise 4.3.45. Show that the rational links associated to the sequences
[[2 3 4]] and [[4 3 2]] are equivalent but that the rational tangles associated
to these sequences are not equivalent.

Exercise 4.3.46. Show that for any nonzero integers a, b, and c, the rational link $[[a\ b\ c]]$ is equivalent to the rational link $[[c\ b\ a]]$. (Challenge: Can this result be generalized to rational "palindromic" sequences with more than three terms, i.e., links of the form $[[a_1\ a_2\ ...\ a_n]]$ and $[[a_n\ ...\ a_2\ a_1]]$?)

Chapter 5

Combinatorial Knot Invariants

We discussed the idea of a knot or link invariant in Section 2.4 as a means of proving that two links are *not* equivalent. In Chapter 5, we will consider some examples as well as a nonexample of knot and link invariants.

5.1   The Writhe of a Diagram

We begin our exploration by recalling the notion of the sign of a crossing from Definition 3.4.6, pictured again in Figure 5.1.1 below.

Figure 5.1.1: Negative and positive crossings.

-1 crossing   +1 crossing

Definition 5.1.1. The **writhe** $w(D)$ of a link diagram D is the sum of the signs of the crossings in the diagram.

Exercise 5.1.2. Choose an orientation for each of the knot diagrams in Figure 5.1.2, and then compute the writhe. Once you've done this, change all orientations and compute the writhes of the oriented knot diagrams with reverse orientations.

Figure 5.1.2: Determine the writhe of these knots.

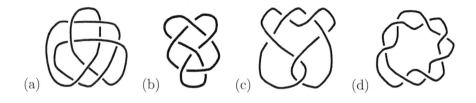

(a)   (b)   (c)   (d)

Exercise 5.1.3. Draw a picture of your favorite knot. Give it an orientation and compute the writhe. Now, give your diagram the opposite orientation. Compute the writhe of this diagram. What do you notice? Formulate and prove a conjecture about the relationship between the writhe of an oriented knot diagram D and its reverse $\bar{D}$.

Exercise 5.1.4. 1. Draw a picture of your favorite 2-component link. Give it an orientation and compute the writhe.

2. Now, give *one of the components* of your diagram the opposite orientation. Compute the writhe of this diagram. What do you notice?

3. Now, change the orientation of the second component and compute the writhe of the resulting diagram. What do you notice?

4. Formulate and prove a conjecture about the relationship between the writhe of an oriented link diagram D and the writhe of link diagrams obtained from D by reversing the orientation of (a) one component or (b) both components of the diagram.

Exercise 5.1.5. Consider the torus knot $T_{2,n}$, where n is a positive odd number. What is the writhe of $T_{2,n}$?

Exercise 5.1.6. Consider the standard rational link diagram D with Conway notation $[[a \ b \ c]]$. What is the writhe of D?

Let's investigate whether or not the writhe is a link invariant. Recall from Chapter 2 that two diagrams represent the same link if and only if the two diagrams are equivalent by a sequence of Reidemeister moves and planar isotopies. Therefore, if we can prove that the application of the Reidemeister moves doesn't change the writhe, then we will have shown that the writhe is an invariant. We'll start our investigation with an analysis of the writhe for a Reidemeister 2 move.

Suppose that a Reidemeister 2 move or its inverse is applied to a link diagram. We will show that the local portion of the link diagram contributes the same value to the writhe before and after the move is performed.

Consider Figure 5.1.3 and observe that the two crossings are signed $+1$ and -1, regardless of how the arcs are oriented. Thus these two crossings contribute a zero sum to the writhe of the diagram. After the Reidemeister 2 move is applied, both crossings are removed. Thus, locally, this portion of the diagram still contributes zero to the writhe. Therefore the writhe is invariant under a Reidemeister 2 move.

Exercise 5.1.7. Finish the investigation into whether or not the writhe of an oriented link diagram is an invariant by considering the following questions.

Figure 5.1.3: Before and after a Reidemeister 2 move.

1. Is the writhe invariant under the Reidemeister 3 move? (To answer this, consider the sums of the signs of the crossings in a link diagram just before and just after a Reidemeister 3 move has been performed.)

2. Is the writhe invariant under the Reidemeister 1 move?

3. What can you conclude? Is the writhe a link invariant? In other words, is the sum of the signs of the crossings in a link diagram preserved by *all* Reidemeister moves?

At this point, you may be wondering how useful the writhe is. We will discover in Chapter 6 that the writhe, given the way it behaves under the Reidemeister moves, can help us to define one of the most widely used knot and link invariants. In addition, one restriction of the writhe provides a simple, but useful, virtual knot invariant in Chapter 8. For now, we will investigate another restriction of the writhe for links that yields a fundamental link invariant.

5.2   The Linking Number

We now have a well-defined notion of the sign of a crossing in a link diagram, but how can we harness this sign convention to develop link invariants? The answer is quite simple. For a two-component oriented link, we define a link invariant called the *linking number*.

Definition 5.2.1. Let J and K be two components of an oriented link diagram. The **linking number** of the link formed by J and K (ignoring any other components the link may have), denoted $\mathrm{lk}(K, J)$, is half the sum of the signs of the crossings where J and K cross.

Exercise 5.2.2. Find at least one pair of link diagrams from Table 1.1 that do not have the same linking number.

Exercise 5.2.3. Show that the oriented Whitehead link, shown in Figure 5.2.1, has linking number 0.

Figure 5.2.1: The oriented Whitehead link.

Exercise 5.2.4. Construct three examples of link diagrams with three different nonzero linking numbers.

Next we prove that the linking number is indeed a link invariant.

Theorem 5.2.5. *If D and D' are two diagrams of a two-component link L, then the linking numbers of D and D' are equal.*

Exercise 5.2.6. Prove Theorem 5.2.5. (Hint: Show that the linking number is unchanged by the application of R1, R2, and R3 moves.)

Exercise 5.2.7. Consider the torus link $T_{2,n}$ where n is a positive even number. In this case, $T_{2,n}$ is a two-component link. What is the linking number of $T_{2,n}$ if both components are oriented in the same direction? What if the two components are oppositely oriented?

Exercise 5.2.8. Find a family of two-component links (other than the $T_{2,n}$ example) such that, for every integer n, there is exactly one member of the family with linking number n.

The definition of linking number makes it clear that $lk(J, K)$ and $lk(K, J)$ are equal. However, it isn't clear why the linking number is always an integer (rather than sometimes equal to half an integer). We will prove that the linking number is indeed always an integer through the following sequence of lemmas.

First, note that the sum in the definition of the linking number above can be split into the sum of the signs of the crossings where the K component crosses over the J component, and the sum of the signs of the crossings where the J component crosses over the K component. Let's call these subsums the K **over** J **sum** and the J **over** K **sum** and denote them by $\sum_{K/J}$ and $\sum_{J/K}$, respectively. Using our sum notation, we have the following.

$$
\begin{aligned}
\mathrm{lk}(J, K) &= \frac{1}{2}\left[\sum_{c,\ \text{a crossing of } J \text{ with } K} \mathrm{sign}(c)\right] \\
&= \frac{1}{2}\left[\sum_{c,\ \text{a crossing } K \text{ over } J} \mathrm{sign}(c) + \sum_{c,\ \text{a crossing } J \text{ over } K} \mathrm{sign}(c)\right] \\
&= \frac{1}{2}\left[\sum_{K/J} + \sum_{J/K}\right]
\end{aligned}
$$

Our first lemma states that the two 'oversums' defined above are unchanged by Reidemeister moves. In other words, both $\sum_{K/J}$ and $\sum_{J/K}$ are invariants in their own rights.

Lemma 5.2.9. *For a link with two components K and J,*
(i) the K over J sum is unchanged by Reidemeister moves, and
(ii) the J over K sum is unchanged by Reidemeister moves.

Exercise 5.2.10. Prove part (i) of Lemma 5.2.9. Part (ii) follows by the same argument.

Next, we investigate how changing a crossing between two components of an oriented link diagram impacts the *difference* between the two oversums.

Lemma 5.2.11. *Let K and J be the components of a link diagram D. Suppose one crossing of component J with component K is changed, resulting in a new diagram D'. Then the difference $\sum_{K/J} - \sum_{J/K}$ does not change value, as D is changed to D'.*

The following lemma looks at a special case. Suppose all the crossings of K with J are crossings of K over J. What would the link look like in this case? Are the two components of the link actually linked?

Lemma 5.2.12. *If the crossings in an oriented link diagram are such that K always passes over J, then the difference $\sum_{K/J} - \sum_{J/K}$ is zero.*

Exercise 5.2.13. Prove Lemma 5.2.12. (Hint: Notice that this link can be deformed so that K and J have disjoint projections.)

Theorem 5.2.14. *The linking number is always an integer.*

Exercise 5.2.15. Prove Theorem 5.2.14. (Hint: Put the previous two lemmas together to make conclusions about an arbitrary link with two components. Start with a diagram D of an arbitrary link of two components J and K. Consider the diagram D' identical to D except that all crossings of K with J are crossings of K over J. By Lemma 5.2.12, what does the difference $\sum_{K/J} - \sum_{J/K}$ equal? Now change D' back to D, one crossing at a time and use Lemma 5.2.11.)

Notice that the argument in Exercise 5.2.15 also proves that the linking number can be defined via either of two oversums, i.e.,

$$\mathrm{lk}(J, K) = \sum_{J/K} = \sum_{K/J}.$$

5.3   Tricolorability

We just explored an example of a quantity that *fails* to be an oriented knot or link invariant as well as a quantity that *is* an invariant of oriented links with two components. We don't yet have an example, however, of an invariant that is defined for knots. So let's explore the idea of a knot coloring. The colorability of a diagram is a new genre of invariant that can be explored both for knots and for links.

Definition 5.3.1. A knot diagram is called **tricolorable** if each arc in the diagram can be drawn using one of three colors, say red (R), yellow (Y), and blue (B), in such a way that the following two conditions hold.
1) At least two colors are used in the diagram.
2) At each crossing, either all arcs are colored the same or all arcs are different colors.

Exercise 5.3.2. (a) Finish assigning colors to the black arcs of the diagram in Figure 5.3.1 to give a valid tricoloring of the diagram.
(b) Determine which diagrams in Table 1.1 are tricolorable.

Figure 5.3.1: An example of a partial tricoloring of a knot diagram.

Exercise 5.3.3. Which of the diagrams in Figure 5.1.2 are tricolorable?

As mentioned above, tricolorability is a link invariant. We prove this by showing it is impossible for one diagram of a knot K to be tricolorable while another diagram of the same knot is not tricolorable.

Theorem 5.3.4. *If a given diagram of a knot, K, is tricolorable, then every diagram of K is tricolorable.*

Theorem 5.3.4 ensures that the following definition makes sense and gives our first example of a knot invariant.

Definition 5.3.5. A knot is called **tricolorable** if its diagrams are tricolorable.

Exercise 5.3.6. Prove Theorem 5.3.4. (Hint: Show that the tricolorability of a diagram is unchanged by the application of R1, R2, and R3 moves.)

Exercise 5.3.7. Prove that the trefoil knot is not equivalent to the unknot.

Exercise 5.3.8. For which values of n is the torus knot $T_{2,n}$ tricolorable?

Exercise 5.3.9. Explore the tricolorability of pretzel links. Make a conjecture and prove it.

Exercise 5.3.10. Are either of the closed braids in Figure 3.4.5 tricolorable?

5.4 A Generalization of Tricolorability

In Section 5.3, we saw how tricolorability is used to distinguish between knots. In this section, we will generalize this idea to a new invariant that uses more than three colors. Instead of adding more colors such as magenta, chartreuse, burnt sienna, and the like, we think of *numbers* as being 'colors.' That is, we label the strands in our diagram with numbers from the set $L_n = \{0, 1, 2, \ldots, n-1\}$. For tricoloring, in particular, we use numbers in the set $\{0, 1, 2\}$ instead of our red, yellow, and blue palette.

To define generalized n-colorability we mimic the first condition in the definition of tricolorability that at least two labels from L_n are used. In generalizing the second half of the definition, however, we have more freedom. One way to interpret the condition that 'at each crossing, either all arcs are colored the same or all arcs are different colors' is to say that once two of the three strands in a crossing have been labeled, say with labels a and b, then the label of the third strand should be determined by some formula depending on a and b. The simplest type of formula would be a linear formula, such as $c = Xa + Yb$, where c is the label of the third strand and X and Y are coefficients selected to ensure that n-colorability is a well-defined invariant.

Figure 5.4.1: At a given crossing, the strands are colored with numbers a, b, c from L_n.

The next exercise investigates the values of X and Y in the linear relationship $c = Xa + Yb$.

Exercise 5.4.1. Consider the labels a, b, c of the strands involved in a crossing as shown in Figure 5.4.1.

1. Notice that the color a is on the overstrand, while b and c are on the understrands. If a and b are known, the linear formula $c = Xa + Yb$ should give the value of c. However, by symmetry, if a and c are

known, the formula $b = Xa + Yc$ should give the value of b. Use these two equations to find an integer value of Y.

2. Now, plug the integer that you found for Y above into the equation $c = Xa + Yb$ so that we can determine what integer X ought to be. Note that the resulting linear equation needs to be satisfied if $a = b = c$ since monochromatic colorings of crossings are required for R1 invariance of colorability. Use this constraint to determine what X must be.

Another constraint on the linear formula $c = Xa + Yb$ is that c should be one of the numbers from $L_n = \{0, 1, 2, \ldots, n-1\}$. This next exercise investigates what can be done to address this issue.

Exercise 5.4.2. Return to the previous definition of tricolorability, but use the 'colors' 0, 1, and 2 instead of red, yellow, and blue. Consider the six crossings in Figure 5.4.2.

1. For each crossing in Figure 5.4.2, the definition of tricoloring requires the value of c to be the color that is not used on the other two strands. Determine the value of c that is forced by this constraint for each crossing.

2. In each of the six crossings, use the linear relationship $c = Xa + Yb$ with the values of X and Y found in Exercise 5.4.1 to determine the integer that c must equal.

3. Compare the results of the previous two tasks. How can c be viewed so that the value of c found using our original tricoloring definition coincides with the requirements forced by the linear formula? (Hint: Modular arithmetic is helpful here.)

Exercises 5.4.1 and 5.4.2 lead to one possible generalization of coloring with more than three colors, as is given in the following definition.

Definition 5.4.3. Let p be an odd prime. A knot diagram is **p-colorable** if each arc of the diagram can be labeled with an integer from 0 to $p-1$ such that:
1) At least two labels are distinct.
2) At each crossing, the relation $2x - y - z \equiv 0 \pmod{p}$ holds, where x is the label on the overstrand and y and z are the other two labels.

Figure 5.4.2: A figure of six crossings labeled from the set $L_3 = \{0, 1, 2\}$. In each crossing, two of the three strands have been colored and one strand, labeled c, is to be determined.

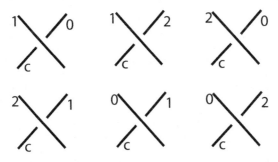

Exercise 5.4.4. Figure 5.4.3 illustrates a partial 7-coloring of a knot. Check that each crossing has a valid labeling and determine the values of a and b.

Figure 5.4.3: Complete this partial 7-coloring.

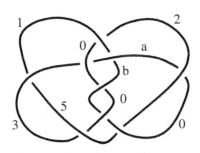

The following is a generalization of Theorem 5.3.4.

Theorem 5.4.5. *If a given diagram of a knot K is p-colorable, then every diagram of K is p-colorable. Thus, p-colorability is a knot invariant.*

To prove Theorem 5.4.5, let D be a p-colorable diagram of a knot K. Recall that any two diagrams of K are related via a sequence of

Reidemeister moves and planar isotopies. Since planar isotopies change only the shape of an arc and do not change the crossings, it suffices to show that after the application of a Reidemeister move, the resulting diagram of K remains p-colorable.

We begin by applying an R2 move to the diagram D and show that the resulting diagram D' is also p-colorable. There are two possibilities we need to consider: (1) D' contains the two additional crossings created by the R2 move; or (2) D contains the two crossings that are removed by the R2 move. We consider case (1) first.

Suppose two arcs in a p-coloring of D are colored x and y as indicated on the left of Figure 5.4.4. After the R2 move, the diagram D' inherits the valid coloring of D on all arcs not depicted in Figure 5.4.4. This means that the arcs emanating into the rest of the diagram for D' must remain colored x and y to ensure the coloring remains valid elsewhere in the diagram. As D was colored using more than two labels, D' also uses more than two labels. Therefore, a valid coloring exists for D' if and only if there exists a numerical color for the arc c that results in a valid coloring on the two new crossings created by the R2 move.

Exercise 5.4.6. Determine the numerical color of arc c in Figure 5.4.4 so that the two crossings adjacent to arc c have valid p-colorings. (Hint: The numerical color for c will depend on x and y.)

Figure 5.4.4: A p-coloring before and after an R2 move.

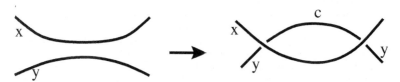

Exercise 5.4.7. Finish the proof of Theorem 5.4.5. That is, complete the proof for the R2 move, case (2), and complete the proofs for the application of R3 and R1 moves.

Exercise 5.4.8. Determine which knots with six or fewer crossings can be 5-colored. For each example, exhibit a 5-coloring. (Figure 2.3.5 contains diagrams of all knots with six or fewer crossings.)

Exercise 5.4.9. For which primes p can the trefoil knot diagram be p-colored? (Hint: Label the three strands x, y, and z, and then write out the system of equations that must be satisfied by a valid coloring. For which primes p can you solve this system?)

Exercise 5.4.10. The p-coloring invariant can be used to distinguish between knots. The knot 8_{16} is both 5- and 7-colorable. However the knots 7_1 and 4_1 are not. Prove this. Diagrams of the knots $8_{16}, 7_1$, and 4_1 are given in Figure 5.4.5.

Figure 5.4.5: The knots 8_{16}, 7_1, and 4_1.

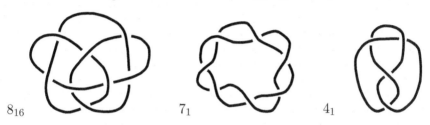

8_{16} $\qquad$ 7_1 $\qquad$ 4_1

5.5   Matrices, Colorings & Determinants

In this section, we will use linear algebra to simplify the problem of finding a p-coloring of a diagram or proving that no such coloring exists. We will also find a new knot invariant along the way.

To translate the problem of coloring a knot diagram into a linear algebra problem, we begin by denoting the color of each arc by a variable. If the knot diagram has n arcs, we color them $x_1, x_2, x_3, \ldots, x_n$. For each crossing in the diagram, we can write down a linear equation relating these variables as follows. If arc x_i passes over a crossing where x_j and x_k are the arcs that come together to form the understrand of the crossing, then we write

$$2x_i - x_j - x_k \equiv 0 \pmod{p}.$$

A p-coloring of the knot exists if there is a solution, $\vec{x}$, to this system of linear equations such that the entries of $\vec{x}$ are not all equal.

Now the problem of finding a p-coloring has been reduced to solving a system of linear equations. As is typical in linear algebra, we will write the

Figure 5.5.1: The diagram of a square knot that contains six arcs.

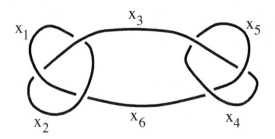

system of linear equations as a matrix equation. Let's work through an example with the **square knot** pictured in Figure 5.5.1. Using the given labeling, the corresponding system of mod p equations is listed in Figure 5.5.2.

Figure 5.5.2: The system of equations from the square knot with labels from Figure 5.5.1.

$$\begin{aligned}
2x_1 - x_2 - x_3 &\equiv 0 \\
-x_1 + 2x_2 - x_6 &\equiv 0 \\
-x_1 - x_2 + 2x_3 &\equiv 0 \\
2x_3 - x_4 - x_5 &\equiv 0 \\
2x_4 - x_5 - x_6 &\equiv 0 \\
?? &\equiv 0
\end{aligned}$$

Exercise 5.5.1. Determine the missing equation in Figure 5.5.2.

This system of equations is written in matrix form in Figure 5.5.3.

Notice that there is a one-to-one relationship between the arcs and the crossings in a knot diagram. One way to see why this is true is to orient the knot and observe that each arc in the diagram emanates from a unique crossing. Therefore, in general, a diagram with n crossings will result in an $n \times n$ matrix. We call this matrix the **crossing-arc matrix** associated to the knot diagram D, denoted by A_D. Notice that each row of the crossing-arc matrix is associated to one of the crossings. Hence, every row contains the entries -1, -1, and 2 and all remaining entries are zero (except in the case where the crossing is an R1 twist, in which case the nonzero entries are -1 and 1). On the other hand, each column of the crossing-arc matrix is associated to one arc of the diagram. Hence, for a given arc, the corresponding column contains two -1 entries for the two crossings involving the arc's endpoints. In addition, for each time the arc passes over a crossing, the column contains a 2. The remaining entries of the column are 0 (except in the case when the arc ends in an R1 twist, in which case the column includes an entry of 1).

Figure 5.5.3: The matrix equation from the system of equations in Figure 5.5.2.

$$
A_D \vec{x} =
\begin{bmatrix}
2 & -1 & -1 & 0 & 0 & 0 \\
-1 & 2 & 0 & 0 & 0 & -1 \\
-1 & -1 & 2 & 0 & 0 & 0 \\
0 & 0 & 2 & -1 & -1 & 0 \\
0 & 0 & 0 & 2 & -1 & -1 \\
a & b & c & d & e & f
\end{bmatrix}
\begin{bmatrix}
x_1 \\
x_2 \\
x_3 \\
x_4 \\
x_5 \\
x_6
\end{bmatrix}
\equiv
\begin{bmatrix}
0 \\
0 \\
0 \\
0 \\
0 \\
0
\end{bmatrix}
$$

Exercise 5.5.2. Fill in the bottom row of the crossing-arc matrix in Figure 5.5.3. Determine the values of a, b, c, d, e, and f.

Standard techniques of linear algebra apply to solving systems of equations mod p. (Formally, for a prime number p, the integers mod p form a field.) Unfortunately, the added condition that at least two of the x_i's differ introduces a few subtleties that need to be addressed before general results can be presented.

We begin with a few observations. First, note that setting each $x_i = 1$ is a solution to the system of equations in Figure 5.5.3. This is true in general because the matrix always consists of rows whose nonzero entries (namely, 2, -1, and -1) sum to zero. (Note, however, that a solution of all 1's is not a solution to the coloring problem because we need at least two distinct entries in the vector of colors.) Second, recall that for any matrix equation $A\vec{x} = \vec{0}$, if $\vec{w}$ and $\vec{v}$ are two solutions and k is any constant, then $\vec{w} + \vec{v}$ and $k\vec{v}$ are also solutions.

These two observations imply that if there is a solution $\vec{w} = (w_1, w_2, \ldots, w_n)$ with not all entries equal, then there is a solution

$\vec{v} = (v_1, v_2, \ldots, v_n)$ with $v_n = 0$, i.e.,

$$\vec{v} = \vec{w} + (-w_n)\vec{1},$$

where $\vec{1}$ denotes an $n \times 1$ vector with all entries equal to 1.

Exercise 5.5.3. Prove that if $\vec{w} = (w_1, w_2, \ldots, w_n)$ is a solution to $A\vec{x} = \vec{0}$ with not all entries equal, then $\vec{v} = \vec{w} + (-w_n)\vec{1}$ is another solution such that $v_n = 0$ and not all entries are equal.

Thus, if for a knot K there is a valid p-coloring, then there is also valid coloring with $v_n = 0$. Observe what this implies about the crossing-arc matrix. A solution with not all entries equal corresponds to a nonzero solution to the system of equations determined by the original matrix with its last column deleted. Looking back at the matrix in Figure 5.5.3, this discussion implies that to find a p-coloring of the square knot, we want to find a *nonzero* solution of the matrix equation in Figure 5.5.4.

Figure 5.5.4: The crossing-arc matrix equation from Figure 5.5.3 with the sixth column and variable x_6 both removed.

$$\begin{bmatrix} 2 & -1 & -1 & 0 & 0 \\ -1 & 2 & 0 & 0 & 0 \\ -1 & -1 & 2 & 0 & 0 \\ 0 & 0 & 2 & -1 & -1 \\ 0 & 0 & 0 & 2 & -1 \\ 0 & 0 & -1 & -1 & 2 \end{bmatrix} \begin{bmatrix} x_1 \\ x_2 \\ x_3 \\ x_4 \\ x_5 \end{bmatrix} \equiv \begin{bmatrix} 0 \\ 0 \\ 0 \\ 0 \\ 0 \\ 0 \end{bmatrix}$$

Recall, from linear algebra, that six vectors in $\mathbb{R}^5$ must be linearly dependent. Hence, the six row vectors in the matrix from Figure 5.5.4 must be linearly dependent. Therefore, one of the six equations contributing to the matrix equation in Figure 5.5.4 is unnecessary, so it is possible to reduce the system down to five equations and five unknowns. In general, the same argument can be applied to reduce the $n \times (n-1)$ matrix equation associated to a knot with n crossings to a square $(n-1) \times (n-1)$ matrix equation. Which equation can we eliminate? In other words, which row of the crossing-arc matrix can we remove?

Let the rows of the square knot's crossing-arc matrix be denoted by $\vec{r_1}, \vec{r_2}, \ldots, \vec{r_6}$. If, for example, we can show that the sixth row is a linear combination of rows 1 through 5, then we can delete the bottom row of the crossing-arc matrix. We will show that this is indeed the case by creating a specific linear combination of the rows, using only coefficients $c_i = \pm 1$, such that the linear combination below, sums to the zero vector.

$$c_1 \vec{r_1} + c_2 \vec{r_2} + \cdots + c_6 \vec{r_6} = \vec{0}$$

This will prove linear dependence of the rows. As row 6 will have either a 1 or -1 for a coefficient, this proves it is dependent on the other five rows and thus the sixth row can be eliminated.

Before determining the coefficients c_i, we must define a *checkerboard coloring* of a link diagram.

Definition 5.5.4. Given a link diagram, D, a **checkerboard coloring** of D, is a coloring of the regions of the diagram using two colors, for example the colors green and white, such that the green regions are adjacent only to white regions, and the white regions are adjacent only to green regions. (Sometimes, we forgo the use of specific color names and use 'shaded' and 'unshaded' to refer to the two colors for the regions.)

Every link diagram can be checkerboard colored in two ways. For examples, see Figure 5.5.5.

Figure 5.5.5: Two checkerboard colorings of a link.

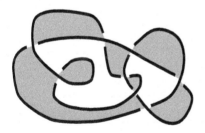 

To determine which coefficients should be multiplied by -1 in our linear combination, we apply the following clever algorithm. First, orient the knot. At each crossing in the diagram put a dot to the right of the overstrand just before entering the crossing. Now, checkerboard color the diagram of the knot so that the color of the unbounded region of the plane is white or unshaded. If the dot for a crossing lies in a shaded region, give the row of the matrix corresponding to that crossing a coefficient of -1, or else use a coefficient of 1. We call this algorithm the **dot-checkerboard algorithm.**

Figure 5.5.6 shows the square knot with a choice of orientation, the checkerboard coloring of the diagram, and the dot associated to each crossing.

Figure 5.5.6: A dotted, checkerboard colored diagram of the square knot.

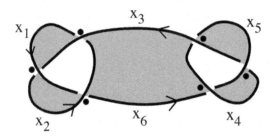

Exercise 5.5.5. Prove that every knot diagram can be checkerboard colored. (Hint: Can the unknot be checkerboard colored? Can every diagram of the unknot be checkerboard colored? Does changing a crossing in a diagram impact the validity of a checkerboard coloring? Also, consider using Exercise 1.2.4.)

Exercise 5.5.6. Prove that the choice of orientation does not influence the outcome of the dot-checkerboard algorithm.

As indicated by Figure 5.5.6, the rows associated with the three crossings on the right-hand side of the diagram, that is, rows 4, 5, and 6, have coefficients of -1. The resulting linear combination now sums to the zero vector.

Exercise 5.5.7. Suppose $\vec{r}_1, \ldots, \vec{r}_6$ are the row vectors from the crossing-arc matrix associated to the square knot discussed above. Check the arithmetic in the following equality.

$$\vec{r}_1 + \vec{r}_2 + \vec{r}_3 - \vec{r}_4 - \vec{r}_5 - \vec{r}_6 = \vec{0}$$

This proves $\vec{r}_6$ is linearly dependent on rows 1 through 5, which implies the bottom row of the crossing-arc matrix can be eliminated without changing the solution space of the matrix equation. Removing the bottom row, we now have a 5×5 altered crossing-arc matrix, denoted $\tilde{A}_D$, as shown in the matrix equation in Figure 5.5.7.

Figure 5.5.7: The altered crossing-arc matrix equation. A nontrivial solution of this equation corresponds to a valid p-coloring of the knot diagram.

$$\tilde{A}_D \vec{x} = \begin{bmatrix} 2 & -1 & -1 & 0 & 0 \\ -1 & 2 & 0 & 0 & 0 \\ -1 & -1 & 2 & 0 & 0 \\ 0 & 0 & 2 & -1 & -1 \\ 0 & 0 & 0 & 2 & -1 \end{bmatrix} \begin{bmatrix} x_1 \\ x_2 \\ x_3 \\ x_4 \\ x_5 \end{bmatrix} \equiv \begin{bmatrix} 0 \\ 0 \\ 0 \\ 0 \\ 0 \end{bmatrix}$$

We proceed by applying methods from linear algebra to the associated square matrix to study whether or not a knot diagram has a p-coloring. Recall, if A is a square matrix, then the equation $A\vec{x} = \vec{0}$ has a nontrivial solution if and only if $\det(A) = 0$, or, since we are working mod p, if and only if the determinant is divisible by p. The 5×5 matrix above has determinant 9. Since 9 is divisible by 3, the knot in Figure 5.5.1 can be 3-colored. Since 9 has no other prime divisors, this knot cannot be p-colored for any prime p other than 3.

The following theorem summarizes the discussion above.

Theorem 5.5.8. *Let A_D denote a crossing-arc matrix associated to a knot diagram, D, with n arcs. Deleting any one column and any one row of A_D yields a new matrix $\tilde{A}_D$. The knot diagram, D, can be p-colored if and only if the matrix equation $\tilde{A}_D \vec{x} = \vec{0}$ has a nontrivial solution mod p.*

Exercise 5.5.9. In the example of the square knot, the last column and the last row of the crossing-arc matrix A were deleted. However, Theorem 5.5.8 states that any one row and any one column of A can be deleted. Explain why it does not matter which row and which column are deleted. (Hint: For the proof that any column can be deleted, study the solution to Exercise 5.5.3. For the proof that any row can be deleted, notice the dot-checkerboard algorithm gives a linear combination of all rows in which every row has a nonzero coefficient.)

Recall that we made a choice when assigning the labels $x_1, x_2, \ldots, x_n$ to the arcs in our diagram. Changing this assignment of labels will result in a different crossing-arc matrix. It is not difficult to see, however, that the process above can be applied and will yield the same result. One way to see this is to return to viewing the crossing-arc matrix equation as a system of linear equations. Notice that the *existence of a solution* to the system of equations has no bearing on the names of the variables used within the system. So the assignment of the labels to the arcs in the knot diagram doesn't influence the outcome.

To complete the proof of Theorem 5.5.8, it suffices to show why the dot-checkerboard algorithm always gives a linear combination of the rows of the crossing-arc matrix that sums to the zero vector.

Exercise 5.5.10. Given an arbitrary knot diagram, D, with n crossings, prove that the dot-checkerboard algorithm described above results in a linear combination of the row vectors of the crossing-arc matrix that sum to the zero vector. (Hint: Analyze an arbitrary arc x_i in your knot diagram. There are several possibilities to consider for x_i. First, assume this arc never passes over a crossing. (Such an arc doesn't exist in an alternating knot diagram, but it may exist in an arbitrary diagram.) In this case, what are the entries of the ith column? In the linear combination produced by the dot-checkerboard algorithm, do these entries sum to zero? Next, analyze the arc x_i, assuming that this arc passes over exactly one crossing. What are the entries of the ith column of the matrix? In the linear combination produced by the dot-checkerboard algorithm, do these entries sum to zero? Continue your analysis of arc x_i, assuming it passes over two, three, or four crossings. Once you determine a pattern, explain what happens in general.)

Now that we have a way of translating the p-coloring question into a question about the existence of a nontrivial solution of the matrix equation $\tilde{A}_D \vec{x} = \vec{0}$, let's recall the powerful relationship between a matrix equation and the determinant of the matrix while reviewing some important ideas from linear algebra.

Linear Algebra Review, Fact 1. A nontrivial solution to $A\vec{x} = \vec{0}$ exists if and only if A is not invertible, and A is not invertible if and only if $\det(A) = 0$. These equivalences still hold when considered modulo a prime p.

Perhaps surprisingly, the determinant gives us information about the colorability of the *knot*, not merely the knot diagram D. Thus, the determinant yeilds another powerful knot invariant. For a knot K and *any* diagram D of K, the **determinant of K** is the absolute value of the determinant of $\tilde{A}_D$, the associated $(n-1) \times (n-1)$ matrix obtained from the crossing-arc matrix by deleting any one column and any one row.

Recall, that the determinant of a 2×2 matrix $A = \begin{bmatrix} a & b \\ c & d \end{bmatrix}$ is given by

$\det(A) = ad - bc$. For larger matrices, determinants can be calculated in several ways, one of which is a recursive process called minor expansion. The ij^{th} minor of an $n \times n$ matrix A, denoted M_{ij}, is the $(n-1) \times (n-1)$ matrix obtained from A by eliminating the ith row and the jth column. Minor expansion can be performed along any row or any column of an $n \times n$ matrix A. Along the ith row, we fix i and use the formula

$$\det(A) = \sum_{j=1}^{n} (-1)^{i+j} a_{ij} \det(M_{ij}).$$

To calculate a determinant along the jth column instead, we fix j and use the formula

$$\det(A) = \sum_{i=1}^{n} (-1)^{i+j} a_{ij} \det(M_{ij}).$$

Notice that this formula is recursive in the sense that one must still calculate the determinant of the minors, M_{ij}. However, the minors are smaller in size than A so if the process is repeated, it will eventually end with the calculation of the determinant of one or more 2×2 matrices.

Here is a quick example showing the first step in calculating the determinant using minor expansion along the fourth row of the matrix M below.

$$\det(M) = \det \begin{bmatrix} 12 & 5 & 0 & -1 \\ 1 & 0 & 6 & 2 \\ 0 & 1 & 10 & 3 \\ 8 & 7 & 0 & 11 \end{bmatrix}$$

$$= (-1)^5 8 \det(M_{41}) + (-1)^6 7 \det(M_{42}) + (-1)^7 0 \det(M_{43}) + (-1)^8 11 \det(M_{44})$$

$$= -8 \det \begin{bmatrix} 5 & 0 & -1 \\ 0 & 6 & 2 \\ 1 & 10 & 3 \end{bmatrix} + 7 \det \begin{bmatrix} 12 & 0 & -1 \\ 1 & 6 & 2 \\ 0 & 10 & 3 \end{bmatrix} + 11 \det \begin{bmatrix} 12 & 5 & 0 \\ 1 & 0 & 6 \\ 0 & 1 & 10 \end{bmatrix}$$

Now there are three 3×3 determinants to calculate to complete the calculation of the determinant of M. Let's look at the calculation of the determinant of the minor M_{44} using minor expansion along the first column.

$$\det \begin{bmatrix} 12 & 5 & 0 \\ 1 & 0 & 6 \\ 0 & 1 & 10 \end{bmatrix}$$

$$= (-1)^2 12 \det \begin{bmatrix} 0 & 6 \\ 1 & 10 \end{bmatrix} + (-1)^3 1 \det \begin{bmatrix} 5 & 0 \\ 1 & 10 \end{bmatrix} + (-1)^4 0 \det \begin{bmatrix} 5 & 0 \\ 0 & 6 \end{bmatrix}$$

$$= 12(-6) + (-1)(50) + 0(30) = -86$$

Exercise 5.5.11. Complete the calculation that we started above for the determinant of the 4×4 matrix M.

Exercise 5.5.12. Find the determinant of each knot in Figure 2.3.5. Using the determinant, find the values of p for which each knot is p-colorable.

To prove that the determinant of a knot gives a well-defined knot invariant, we must show it is invariant under Reidemeister moves. In preparation for this, we recall some important observations and theorems from linear algebra. These results will be useful to us as we complete several upcoming exercises.

First, let's look at a definition. Given an $n \times n$ matrix A, the **elementary matrix that corresponds to a specified row or column operation** on A is the matrix that is obtained by performing the specified row or column operation on the $n \times n$ identity matrix.

Linear Algebra Review, Fact 2. Performing an elementary row (respectively, column) operation on a matrix A is equivalent to multiplying A on the left (respectively, right) by the corresponding elementary matrix.

For instance, the row operation "add 5 times row 1 to row 2" (when applied to a 3×3 matrix) corresponds to the elementary matrix

$$E = \begin{bmatrix} 1 & 0 & 0 \\ 5 & 1 & 0 \\ 0 & 0 & 1 \end{bmatrix}.$$

This correspondence is useful because, as Fact 1 implies, the matrix product EB, for an arbitrary 3×3 matrix B, gives the result of "adding 5 times row 1 of B to row 2 of B" in a much more succinct expression (You should check this product!). Also notice that if we calculate the matrix product in the reverse order, BE, then matrix E performs the column operation "add 5 times column 2 of B to column 1 of B." This is to be expected since E can also be viewed as the result of this exact column operation applied to the 3×3 identity matrix. So, by multiplying the elementary matrix E on the left of B, we perform the *row* operation corresponding to E, and multiplying E on the right of B performs the *column* operation corresponding to E.

Here is an example of this observation that includes several row and column operations applied to a 2×2 matrix A.

Consider the matrix $A = \begin{bmatrix} 1 & 4 \\ 2 & 0 \end{bmatrix}$ and apply the following row and column operations.

- add two times row 1 of A to row 2

- swap columns 1 and 2

- multiply row 2 by $\frac{1}{2}$

- add the negative of row 2 to row 1

- add -3 times column 1 to column 2

Exercise 5.5.13. Prove that the list of row and column operations above transforms A into $B = \begin{bmatrix} 0 & -1 \\ 4 & -10 \end{bmatrix}$.

Applying Linear Algebra Review, Fact 2, the 2×2 matrix A and the list of row and column operations that transform A into B are equivalent to the following matrix product equality.

$$\begin{bmatrix} 1 & -1 \\ 0 & 1 \end{bmatrix}\begin{bmatrix} 1 & 0 \\ 0 & \frac{1}{2} \end{bmatrix}\begin{bmatrix} 1 & 0 \\ 2 & 1 \end{bmatrix}\begin{bmatrix} 1 & 4 \\ 2 & 0 \end{bmatrix}\begin{bmatrix} 0 & 1 \\ 1 & 0 \end{bmatrix}\begin{bmatrix} 1 & -3 \\ 0 & 1 \end{bmatrix} = \begin{bmatrix} 0 & -1 \\ 4 & -10 \end{bmatrix}$$

Exercise 5.5.14. (1) Calculate the matrix product

$$\begin{bmatrix} 1 & -1 \\ 0 & 1 \end{bmatrix} \begin{bmatrix} 1 & 0 \\ 0 & \frac{1}{2} \end{bmatrix} \begin{bmatrix} 1 & 0 \\ 2 & 1 \end{bmatrix} \begin{bmatrix} 1 & 4 \\ 2 & 0 \end{bmatrix} \begin{bmatrix} 0 & 1 \\ 1 & 0 \end{bmatrix} \begin{bmatrix} 1 & -3 \\ 0 & 1 \end{bmatrix} \text{ to show}$$

that it is equal to $\begin{bmatrix} 0 & -1 \\ 4 & -10 \end{bmatrix}$.

(2) Comment on the order of the elementary matrices in this product. Why is the matrix $\begin{bmatrix} 1 & -1 \\ 0 & 1 \end{bmatrix}$ the first in the product? Why is $\begin{bmatrix} 1 & -3 \\ 0 & 1 \end{bmatrix}$ last in this product?

Next, we recall a few more facts from linear algebra. You might also want to review these results by reading any standard linear algebra textbook.

Linear Algebra Review, Fact 3. The determinant is multiplicative. That is, for $k \times k$ matrices A and B, we have the equality $\det(AB) = \det(A)\det(B)$.

Linear Algebra Review, Fact 4.

- If E is an elementary matrix corresponding to adding a multiple of a row/column to another row/column, then $\det(E) = 1$.

- If E is an elementary matrix corresponding to swapping two rows/columns, then $\det(E) = \pm 1$.

- If E is an elementary matrix corresponding to multiplying a row/column by the scalar k, then $\det(E) = k$.

This completes our linear algebra review. Returning to our main goal of this section, we can show that the determinant of a knot remains constant after the application of each of the three Reidemeister moves. To begin the proof of invariance under an R1 move, consider the following notation. Let D be a diagram with n crossings and, hence, n arcs. Select an assignment of the arc labels $x_1, x_2, \ldots, x_n$, as pictured in Figure 5.5.8, so that the R1 move is performed on the arc labeled x_n. Suppose, also, that x_n passes over k crossings on one side of the R1 move and passes over j crossings on the other side of the R1 move (k and/or j may equal 0). After the R1 move, we keep the arc labels constant with one exception: the new arc will be labeled x_{n+1}.

Figure 5.5.8: Notation for the crossing-arc matrix of a knot before and after an R1 move.

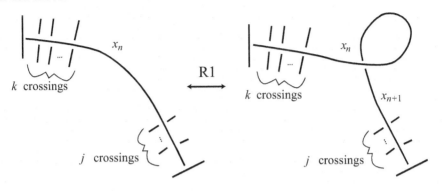

Using the notation in Figure 5.5.8, we have the two crossing-arc matrices shown in Figure 5.5.9.

Figure 5.5.9: The crossing-arc matrices associated with the two diagrams in Figure 5.5.8, with A^* denoting the $n \times (n-1)$ matrix containing the crossing-arc values from arcs $x_1, \ldots, x_{n-1}$.

$$
\begin{array}{c}
x_n \\
\left[\begin{array}{cc}
 & -1 \\
 & 2 \\
A^* & \vdots \\
 & 2 \\
 & 2 \\
 & \vdots \\
 & 2 \\
 & -1 \\
 & 0 \\
 & \vdots \\
 & 0
\end{array} \right]
\begin{array}{l}
\left.\begin{array}{l} \\ \\ \\ \end{array}\right\} \begin{array}{l} k \\ \text{crossings} \end{array} \\
\left.\begin{array}{l} \\ \\ \end{array}\right\} \begin{array}{l} j \\ \text{crossings} \end{array}
\end{array}
\end{array}
\qquad
\xrightarrow{\text{R1}}
\qquad
\begin{array}{c}
x_n \quad x_{n+1} \\
\left[\begin{array}{ccc}
 & -1 & \\
 & 2 & \\
A^* & \vdots & \\
 & 2 & \\
 & 0 & 2 \\
 & \vdots & \vdots \\
 & & 2 \\
 & & -1 \\
 & 0 & 0 \\
 & \vdots & \vdots \\
 & 0 & 0 \\
0 \ldots 0 & 1 & -1
\end{array} \right]
\end{array}
$$

new crossing →

Exercise 5.5.15. Prove that the determinant of a knot is unchanged by the application of an R1 move. (Hint: Use row and column operations and the linear algebra review facts to prove that both matrices in Figure 5.5.9

have the same determinant as the matrix in Figure 5.5.10.)

Figure 5.5.10: Both matrices in Figure 5.5.9 have determinants equal to the determinant of the $n \times n$ matrix below, up to a factor of ± 1.

$$\begin{bmatrix} & & -1 & 0 \\ & & 2 & 0 \\ & & \vdots & \vdots \\ & & 2 & 0 \\ & & 2 & 0 \\ & A* & \vdots & \vdots \\ & & 2 & 0 \\ & & -1 & 0 \\ & & 0 & 0 \\ & & \vdots & \vdots \\ & & 0 & 0 \\ 0 \ldots & & 0 & 1 \end{bmatrix}$$

Exercise 5.5.16. Prove that the determinant of a knot is unchanged by the application of an R2 move. (Hint: Set up your notation with care, as we did in Exercise 5.5.15.)

Exercise 5.5.17. Prove that the determinant of a knot is unchanged by the application of an R3 move.

Chapter 6

Knot Polynomials

6.1   The Alexander Polynomial

The Alexander polynomial, a knot invariant discovered by James W. Alexander in 1923, was the first and only polynomial knot invariant for over 60 years. The polynomial can be calculated by evaluating the determinant of a matrix associated to a diagram of a knot. In addition to picking a diagram of the knot to compute the Alexander polynomial, many choices are made when the matrix is constructed. It is a rather amazing polynomial, but there is much to prove when showing that the construction that produces the polynomial is an invariant of knots.

Let K be a knot with an n crossing knot diagram D. Choose an orientation for D. Label the arcs and the crossings of the diagram each with the integers 1 through n. Create an $n \times n$ matrix with entries in row w determined by the arc labels that meet at crossing w. If, while following the orientation along arc i, we encounter arc j to the left and k to the right of crossing w, then the matrix entries of row w are $1 - t$ in column i, t in column j and -1 in column k. All other entries in row w are zero. If any of i, j, or k happen to be equal, then we put the sum of the entries into the appropriate column.

Notice that this matrix is similar to the matrix we used when calculating the determinant of a knot. However, instead of the row w containing the entries $2, -1, -1$, it contains the linear functions $1 - t$, t, and -1. Observe that when this matrix is evaluated at $t = -1$, we obtain the crossing-arc matrix used in calculating the determinant of the knot. Also notice that, just as with the matrix that is used to compute the determinant of a knot, the column vectors of this matrix sum to the zero vector.

Figure 6.1.1: The Alexander polynomial matrix entries for crossing w.

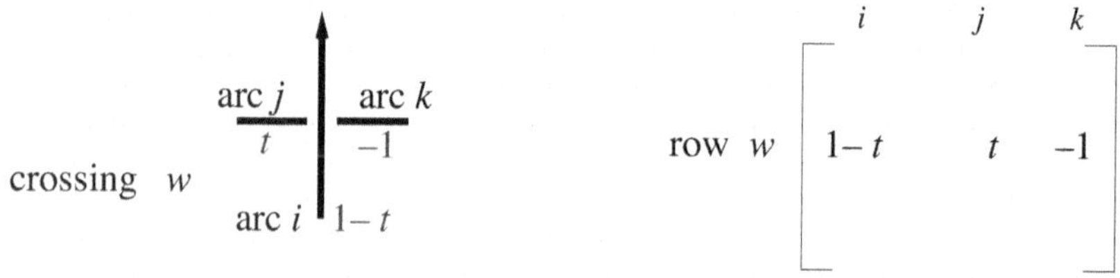

Definition 6.1.1. Given an oriented and labeled diagram D of a knot K, create the matrix M_D whose entries are described in Figure 6.1.1. The $(n - 1) \times (n - 1)$ matrix obtained by removing the bottom row and last

column from M_D is called an **Alexander matrix** of the diagram D. The determinant of an Alexander matrix, denoted by $A_D(t)$, is called an **Alexander polynomial of the diagram** D. The determinant of a 0×0 matrix is defined to be 1, and thus, the Alexander polynomial of the unknot is defined to be 1.

The following examples show that the Alexander polynomial as described is not quite a knot invariant. It depends on the knot diagram, the orientation, and the choices made in the knot labeling. However, we will discover that an Alexander polynomial can be normalized to produce a bonafide knot invariant!

Exercise 6.1.2. To calculate the Alexander polynomial of the figure eight knot diagram shown in Figure 6.1.2, we begin by selecting an orientation, as well as a labeling of the arcs and crossings. One possibility is to make the choices shown in Figure 6.1.2. For this diagram, we label the crossings by the label of the overstrand arc.

1. Find the 4×4 matrix associated with these choices.

2. Find the Alexander matrix associated with these choices.

3. Find the resulting Alexander polynomial of the diagram in Figure 6.1.2.

Figure 6.1.2: Oriented and labeled figure eight knot.

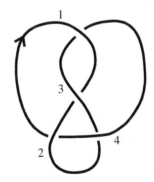

Exercise 6.1.3. Next, consider the figure eight knot shown in Figure 6.1.3, with crossing labels shown. Label each arc according to the crossing that the arc passes over.

1. Find the 5×5 matrix associated with these choices.

2. Find the Alexander matrix associated with these choices.

3. Find the resulting Alexander polynomial of the diagram in Figure 6.1.3.

Figure 6.1.3: Labeled figure eight knot with R1 move performed.

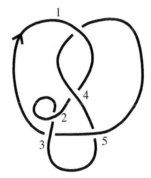

Exercise 6.1.4. (1) Calculate the Alexander polynomial of the left-handed trefoil knot from the knot diagram that has three negative crossings.
(2) Calculate the Alexander polynomial of the right-handed trefoil knot from the knot diagram that has three positive crossings.
(3) Perform an R2 move on the right-handed trefoil, and then calculate the Alexander polynomial for this new diagram.

As we saw in the previous exercises, the Alexander polynomial depends on the choice of the knot diagram and the choices made when labeling the diagram. However, we also observe that the polynomials obtained by making various choices differ by no more than a factor of $\pm t^k$ for some integer k.

Theorem 6.1.5. *Suppose D is an oriented knot diagram and consider two different labelings of the arcs and crossings of D. Denote the two labeled diagrams by D_1 and D_2. The Alexander polynomials of the two labelings satisfy $A_{D_1}(t) = \pm t^k A_{D_2}(t)$ for some integer k.*

Exercise 6.1.6. Prove Theorem 6.1.5. (Hint: A change in labeling of exactly two arcs of D corresponds to swapping the two corresponding columns of the Alexander matrix. Recall that swapping two columns of a matrix is an elementary operation. Find the analogous correspondence for a change in the labels for exactly two crossings. When the corresponding elementary operations are applied to the Alexander matrix, what is the effect on the Alexander polynomial?)

Theorem 6.1.7. *Suppose D and D' are two different diagrams of an oriented knot K. Then the Alexander polynomials of the two diagrams satisfy $A_D(t) = \pm t^k A_{D'}(t)$ for some integer k.*

Together, we will prove Theorem 6.1.7 in the case where D and D' differ by an R3 move. The remaining cases of Theorem 6.1.7 are left as Exercise 6.1.10.

Suppose the oriented diagrams D and D' differ by an $R3$ move, as shown in Figure 6.1.4. Label the arcs for D and D' identically throughout the diagram except in the local region where the R3 move is made. For these local arcs, using the labeling scheme shown in Figure 6.1.4. Observe that the arcs that extend beyond the local regions are consistently labeled before and after the R3 move. Next, we create the two matrices M_D and $M_{D'}$ corresponding to the labeled diagrams before and after the R3 move, as shown in Figures 6.1.5 and 6.1.6. The matrix B^* is identical in both M_D and $M_{D'}$ due to the choice of identical labeling for the portions of diagram D and D' that are not pictured.

Figure 6.1.4: An oriented R3 move with labeled arcs, $a_1, \ldots, a_6$ and labeled crossings C_1, C_2, C_3.

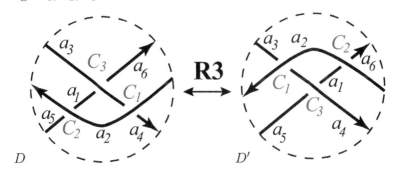

Exercise 6.1.8. Use Figure 6.1.4 to check the matrix entries in the top three rows of M_D and $M_{D'}$, as shown in Figures 6.1.5 and 6.1.6. Explain why column 1, in both matrices, contains only zeros below the third row. Explain why rows $1-3$, for both matrices, contain only zeros in the entries after the sixth column.

Our goal now is to use elementary row and column operations to transform M_D into a form that is similar to $M_{D'}$. Initially, we do not want the

Figure 6.1.5: The matrix M_D associated to diagram in D from Figure 6.1.4. The submatrix B^* contains the matrix entries for the remaining crossings in the diagram.

$$
\begin{array}{c}
\\ C_1 \\ C_2 \\ C_3 \\ \\ \\ \\
\end{array}
\begin{array}{cccccccc}
a_1 & a_2 & a_3 & a_4 & a_5 & a_6 & \text{other arcs} \\
\left[\begin{array}{ccccccc}
0 & 1-t & -1 & t & 0 & 0 & 0 \ldots 0 \\
-1 & 1-t & 0 & 0 & t & 0 & 0 \ldots 0 \\
-1 & 0 & 1-t & 0 & 0 & t & 0 \ldots 0 \\
0 & & & & & & \\
& & & & B^* & & \\
\vdots & & & & & & \\
0 & & & & & &
\end{array}\right]
\end{array}
$$

Figure 6.1.6: The matrix $M_{D'}$ associated to diagram in D' from Figure 6.1.4. The submatrix B^* contains the matrix entries for the remaining crossings in the diagram and is identical to B^* in Figure 6.1.5.

$$
\begin{array}{c}
\\ C_1 \\ C_2 \\ C_3 \\ \\ \\ \\
\end{array}
\begin{array}{cccccccc}
a_1 & a_2 & a_3 & a_4 & a_5 & a_6 & \text{other arcs} \\
\left[\begin{array}{ccccccc}
0 & 1-t & -1 & t & 0 & 0 & 0 \ldots 0 \\
t & 1-t & 0 & 0 & 0 & -1 & 0 \ldots 0 \\
t & 0 & 0 & 1-t & -1 & 0 & 0 \ldots 0 \\
0 & & & & & & \\
\vdots & & & & B^* & & \\
0 & & & & & &
\end{array}\right]
\end{array}
$$

elementary row and column operations to alter B^*, so we restrict to using row operations on the top three rows, and column operations that add a multiple of column 1 to some other column.

Exercise 6.1.9. (a) Apply the following list of row and column operations to the matrix M_D from Figure 6.1.5.

(1) Add $(1 - t)$ times row 1 to row 3.
(2) Add t times column 1 to column 5.
(3) Add t times column 1 to column 6.
(4) Multiply row 2 by t^{-1}.
(5) Multiply row 3 by t^{-1}.
(6) Multiply column 1 by $-t^2$.

Let the matrix resulting from these five operations be denoted by $\tilde{M}_D$.

(b) Observe that $\tilde{M}_D$ is now identical to $M_{D'}$, except in two of the matrix entries.

(c) In the next manipulation, we alter $\tilde{M}_D$ and $M_{D'}$ in an identical way so that any changes to B^* are the same in both matrices.

(7) Swap the second and last columns of $\tilde{M}_D$. Name the result $\tilde{M}'_D$.
(8) Swap the second and last columns of $M_{D'}$. Name the result $M'_{D'}$.

(d) For each matrix manipulation $(1) - (8)$, state the effect it has on the determinant of the matrix.

(e) The Alexander polynomial is obtained by deleting the last column and bottom row of $\tilde{M}'_D$ and $M'_{D'}$ and taking the determinant of the resulting matrices. Prove that the Alexander polynomials of D and D' differ by no more than a factor of $\pm t^k$.

Exercise 6.1.10. Prove the remaining cases of Theorem 6.1.7. (Hint: You must prove the result holds if D and D' differ by a single Reidemeister move. Use Figure 2.3.6 to prove the result holds for a generating set of oriented Reidemeister moves. For each Reidemeister move, take special care to set up your notation and the matrices M_D and $M_{D'}$. Find elementary operations that can be applied to these two matrices that show they lead to Alexander polynomials that differ by no more than a factor of $\pm t^k$.)

Next, we investigate the influence of the choice of orientation on the calculation of the Alexander polynomial. We will see that the coefficients of the Alexander polynomial exhibit a special symmetry.

Exercise 6.1.11. Let D denote the diagram of the figure eight knot with labels and orientation as in Figure 6.1.2. Find the Alexander polynomial of the reverse, $\bar{D}$. Show that $A_{\bar{D}}(t^{-1}) = \pm t^k A_D(t)$ for some integer k.

Theorem 6.1.12. *Suppose D is an oriented and labeled diagram and its reverse is $\bar{D}$. The Alexander polynomial of the diagram and its reverse satisfy $A_{\bar{D}}(t^{-1}) = \pm t^j A_D(t)$ for some integer j.*

Exercise 6.1.13. Prove Theorem 6.1.12.

We can now define *the* Alexander polynomial of a knot as a normalized version of the polynomial that we have been calling the Alexander polynomial of a diagram D.

Definition 6.1.14. Given the Alexander polynomial, $A_D(t)$, for a labeled and oriented diagram, D, of a knot, K, factor the polynomial as

$$\pm t^k A_K(t) = A_D(t),$$

for some integer k, so that $A_K(t)$ has a positive constant term. The resulting polynomial, $A_K(t)$, is the **Alexander polynomial of the knot** K.

The polynomial $A_K(t)$ is independent of the choice of labeling by Theorem 6.1.5, is independent of the choice of diagram by Theorem 6.1.7, and is independent of the choice of orientation by Theorem 6.1.12. Therefore, the Alexander polynomial, $A_K(t)$, is an invariant of unoriented knots.

Exercise 6.1.15. Calculate the Alexander polynomial of the seven crossing knot diagram in Figure 6.1.7.

Figure 6.1.7: A knot diagram with seven crossings.

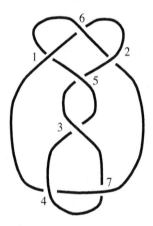

The Alexander polynomial can be used to distinguish many knots from one another, but there are some knots it cannot differentiate.

Exercise 6.1.16. Show that the knot in Figure 6.1.8 has Alexander polynomial 1.

The knot in Figure 6.1.8 is one of two knots with 11 crossings that has a trivial Alexander polynomial. No knots with 10 or fewer crossings has a trivial Alexander polynomial, other than the unknot.

Figure 6.1.8: A nontrivial knot with trivial Alexander polynomial.

6.2   The Kauffman Bracket & Jones Polynomial

Another polynomial, discovered in 1984 by Vaughn F. R. Jones, associates a Laurent polynomial[1] with integer coefficients [16] to every knot or link. Unlike the Alexander polynomial, the Jones polynomial can sometimes distinguish between nonequivalent mirror images of a knot. Indeed, we will see that the Jones polynomial can distinguish between the **left-handed trefoil** and **right-handed trefoil**, that is, the trefoil that can be drawn with three negative crossings and the trefoil that can be drawn with three positive crossings. The Jones polynomial is a stronger invariant than the Alexander polynomial. For example, currently there is no known example of a nontrivial knot K whose Jones polynomial is trivial, whereas the knot in Figure 6.1.8 is such a knot for the Alexander polynomial.

We will define the Jones polynomial of a link L through a recursively-defined function called the Kauffman bracket [20]. The Kauffman bracket is a function, $\langle L \rangle$, that assigns a Laurent polynomial to an unoriented link diagram L. To compute the Kauffman bracket of a link diagram, we provide a way of relating the bracket of a nontrivial link diagram to a pair of brackets of simpler link diagrams. Specifically, the Kauffman bracket of a link diagram, L_n, with n crossings, for $n \geq 1$, is defined to be equal to the value of a certain sum,

$$A\langle L_{n-1}^{+} \rangle + A^{-1}\langle L_{n-1}^{-} \rangle,$$

involving the Kauffman bracket of two associated link diagrams, each with

---

[1]Laurent polynomials are finite polynomial expressions that can include negative powers as well as positive powers of variables.

$n - 1$ crossings. The two simpler link diagrams, L_{n-1}^+ and L_{n-1}^-, are obtained from the original diagram L_n by *smoothing* at a crossing of L_n in both possible ways. In other words, a crossing is replaced by a [0] or [∞] tangle. This is called Rule 3.

We observe that when smoothing a crossing in a link, a disconnected, trivial unknot diagram may arise. If this happens, we can simplify the bracket of the disjoint union $L' \cup U$, where U denotes a crossingless diagram of the unknot. The bracket $\langle L' \cup U \rangle$ is replaced by a multiple of the Kauffman bracket of L', namely $(-A^2 - A^{-2})\langle L' \rangle$. *Voila!* We have Rule 2.

Finally, we define $\langle U \rangle$, the bracket of the unknot diagram with no crossings, to be 1. This base case is called Rule 1.

These three rules are summarized in the following definition.

Definition 6.2.1. The **Kauffman bracket** of a link diagram, L, is a polynomial in integer powers of the variable A, denoted by $\langle L \rangle$, defined by the following three rules.

1. $\langle \bigcirc \rangle = 1$

2. $\langle L' \cup \bigcirc \rangle = (-A^2 - A^{-2})\langle L' \rangle$

3. $\langle \times \rangle = A \langle \,) (\,\rangle + A^{-1} \langle \asymp \rangle$

   $\langle \times \rangle = A^{-1} \langle \,) (\,\rangle + A \langle \asymp \rangle$

Notice that for any link, we can repeatedly apply Rule 3, reducing the number of crossings at each step. We can apply Rule 2 to eliminate disjoint unknotted components. Ultimately, we will be left with a single unknot component to which we can apply Rule 1. In Exercise 6.2.4, we show that the order in which these rules are applied does not affect the final result. Thus, the Kauffman bracket is a well-defined quantity assigned to a link diagram.

First, let's see these rules in action on a very simple pair of examples.

Example 6.2.2. We compute the following bracket polynomials:

$\langle \, \mathcal{CO} \, \rangle$ and $\langle \, \mathcal{OO} \, \rangle$.

$$\langle \; \infty \; \rangle = A\langle \; \bigcirc \;\; \bigcirc \; \rangle + A^{-1}\langle \bigcirc \rangle \qquad \text{(Rule 3)}$$

$$= A(-A^2 - A^{-2})\langle \bigcirc \rangle + A^{-1}\langle \bigcirc \rangle \qquad \text{(Rule 2)}$$

$$= A(-A^2 - A^{-2}) + A^{-1} \qquad \text{(Rule 1)}$$

$$= -A^3$$

$$\langle \; \infty \; \rangle = A^{-1}\langle \; \bigcirc \;\; \bigcirc \; \rangle + A\langle \bigcirc \rangle \qquad \text{(Rule 3)}$$

$$= A^{-1}(-A^2 - A^{-2})\langle \bigcirc \rangle + A\langle \bigcirc \rangle \qquad \text{(Rule 2)}$$

$$= A^{-1}(-A^2 - A^{-2}) + A \qquad \text{(Rule 1)}$$

$$= -A^{-3}$$

Now, let's all try computing the Kauffman bracket!

Exercise 6.2.3. Show that the bracket polynomial of the left-handed trefoil shown in Figure 6.2.1 is $A^7 - A^3 - A^{-5}$. You may use the results of Example 6.2.2 to simplify your calculation, if you like.

Figure 6.2.1: The left-handed trefoil.

Exercise 6.2.4. (a) Show that when Rule 3 for the bracket polynomial is applied to two crossings of a diagram, the order in which it is applied does not affect the result.

(b) Show that when Rules 2 and 3 for the bracket polynomial are applied to a link with a disjoint union of the unknot, the order in which they are applied does not affect the result.

Exercise 6.2.5. Find the effect on the Kauffman bracket of performing an R2 move. In other words, what should X and Y be in Figure 6.2.2?

Figure 6.2.2: The Kauffman bracket of an R2 move.

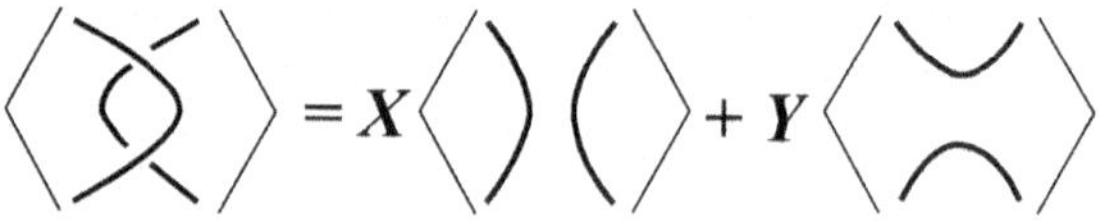

$$\left\langle \text{⤫} \right\rangle = X \left\langle \text{)(} \right\rangle \left\langle \text{()} \right\rangle + Y \left\langle \text{⌣⌢} \right\rangle$$

In Exercise 6.2.5, you proved that the Kauffman bracket is invariant under the R2 move. Indeed, this invariance is how Kauffman determined the terms A^{-1} and $(-A^2 - A^{-2})$ in the definition of the Kauffman bracket. Let's take a look at how the invariance of the Kauffman bracket under R3 moves is a direct result of its invariance under R2 moves as shown in Figure 6.2.3.

Figure 6.2.3: The Kauffman bracket of an R3 move.

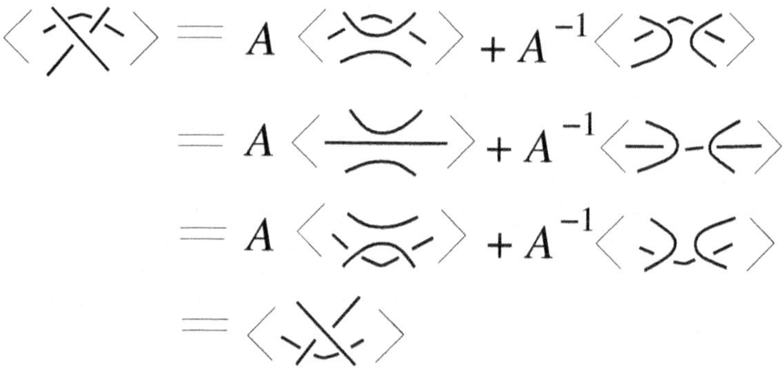

$$\left\langle \text{⤬} \right\rangle = A \left\langle \text{⤬} \right\rangle + A^{-1} \left\langle \text{)(} \right\rangle$$
$$= A \left\langle \text{⤬} \right\rangle + A^{-1} \left\langle \text{)-(} \right\rangle$$
$$= A \left\langle \text{⤬} \right\rangle + A^{-1} \left\langle \text{)(} \right\rangle$$
$$= \left\langle \text{⤬} \right\rangle$$

In order to determine whether or not the Kauffman bracket is a link invariant, it remains to consider the R1 moves. As we saw in Example 6.2.2, there seems to be a problem with this type of Reidemeister moves. Indeed, the Kauffman bracket is *not* invariant under R1 moves.

Figure 6.2.4: The Kauffman bracket of an R1 move.

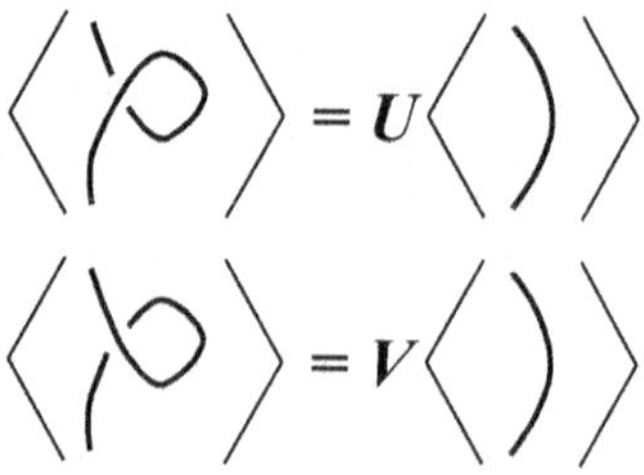

$$\left\langle \text{⟲} \right\rangle = U \left\langle \text{)} \right\rangle$$
$$\left\langle \text{⟳} \right\rangle = V \left\langle \text{)} \right\rangle$$

Exercise 6.2.6. Determine the coefficients U and V of the bracket equations in Figure 6.2.4.

Can you think of another quantity we've seen that is invariant under R2 and R3 moves, but not R1 moves? That's right! The writhe! (Recall Definition 5.1.1.) Because both the writhe and the Kauffman bracket fail to be link invariants in similar ways, the writhe of a link is used to adjust for the lack of invariance of the bracket under an R1 move. Using this clever trick, we can create a link invariant.

Based on our findings in Exercise 6.2.6, we see that multiplying the bracket polynomial $\langle L \rangle$ of an oriented link L by $(-A)^{-3w(L)}$, where $w(L)$ is the writhe of L, will cancel out the noninvariant behavior of the bracket of an R1 move. Using Exercises 5.1.7 and 6.2.5 as well as the computation in Figure 6.2.3, we can prove that the polynomial, $K_L(A)$, called the **Kauffman bracket polynomial** or the **Kauffman polynomial**,

$$K_L(A) = (-A)^{-3w(L)}\langle L \rangle$$

is a link invariant. Notice that for a knot, the invariant does not depend on the choice of orientation, by our result in Exercise 5.1.3. However, for a link of two or more components, the writhe will depend on the choice of orientation of the components. (See Exercise 5.1.4.) This is precisely why we now require L to be oriented.

Exercise 6.2.7. Prove that the Kauffman bracket polynomial is an invariant of oriented links.

At last, we are in a position to define the Jones polynomial.

Definition 6.2.8. The **Jones polynomial** of a link L, denoted by $V_L(t)$, is obtained by making the variable substitution $A = t^{-1/4}$ in the polynomial $K_L(A)$.

Exercise 6.2.9. Using Exercise 6.2.3, determine the Jones polynomial of the left-handed trefoil knot.

Exercise 6.2.10. Calculate the Jones polynomial of the right-handed trefoil knot. Use Exercise 6.2.9 to conclude that the trefoil knot is not equivalent to its mirror image.

Exercise 6.2.11. Choose orientations for the components of the Whitehead link (see Figure 5.2.1) and compute the Jones polynomial. Compute the Jones polynomial of the unlink with two components. Use these results to conclude the Whitehead link is a nontrivial link. (To do your Kauffman polynomial computation, you may prefer to use the diagram of the Whitehead link with five crossings, shown in Table 1.1, example (i).)

6.3   Tait's Conjecture

The discovery of the Jones polynomial led to an explosion of new ideas and results in knot theory. One such result is a proof of one of the renowned Tait Conjectures, which were first proposed in the late 19th century by physicist Peter Guthrie Tait [38]. We alluded to one of the Tait Conjectures in Section 1.3, Exercise 1.3.6. This exercise asks you to play with alternating knots with the goal of conjecturing their crossing number. Several diagrams in Figure 1.3.1 have one or more crossings that can be untwisted, thus reducing the number of crossings in the diagram. The crossings that can be untwisted are called *reducible crossings*. A reducible crossing can be readily identified in a diagram by observing that the four regions adjacent to a reducible crossing are not distinct regions in the diagram, as in Figure 6.3.1.

Definition 6.3.1. A **reducible crossing** is a crossing in a knot diagram such that a single region in the diagram is *twice* adjacent to that crossing. A diagram without any reducible crossings is called a **reduced diagram**.

Figure 6.3.1: An example of a diagram with a reducible crossing. The shaded region below is twice adjacent to the reducible crossing c.

Exercise 6.3.2. Consider the diagrams in Figure 1.3.1. Find all of the crossings in these diagrams that are reducible. Are any of these diagrams reduced?

The idea of the first Tait Conjecture is that once an *alternating* diagram is reduced, the number of crossings in the diagram cannot be decreased.

Theorem 6.3.3. [**Tait's Conjecture**] *If L is a connected and reduced alternating diagram of a link with n crossings, then the crossing number of L is n. Equivalently, any two connected, reduced alternating diagrams of a link L have the same number of crossings.*

As you can see, Tait's Conjecture specifically applies to *connected* link diagrams. Let's look at a definition.

Definition 6.3.4. A link diagram L is called **connected** provided that the corresponding link shadow is a connected subspace of the plane.

We will prove Tait's Conjecture using the Kauffman polynomial. However, rather than using the recursive definition of the bracket given in Definition 6.2.1, we introduce an explicit formula for the Kauffman bracket in terms of smoothed states of a diagram.

Recall the recursive formula for the Kauffman polynomial that reduces evaluating the Kauffman bracket on a diagram with n crossings to evaluating it on a diagram with $n - 1$ crossings.

$$\langle \asymp \rangle = A \langle \rangle\langle \rangle + A^{-1} \langle \underset{\smile}{\frown} \rangle$$

We call the smoothing that is paired with the coefficient A an **A-smoothing** and the smoothing that is paired with the coefficient A^{-1} an **A^{-1}-smoothing** of the crossing. For a link diagram L, the diagram obtained by performing a smoothing of type A or type A^{-1} for every crossing of L is called a **smoothed state** of L.

For a link diagram L and state S of L that contains i A-smoothings and j A^{-1}-smoothings, we define the **bracket of L for the state S** as

$$\langle L|S \rangle = A^i (A^{-1})^j.$$

Figure 6.3.2: The figure eight knot—commonly referred to by its name, 4_1, from the Rolfsen knot table [33]—together with two of its 16 smoothed states. Each crossing in the smoothed state is labeled with an A or A^{-1}, according to its smoothing type.

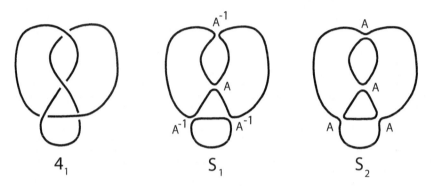

For example, using the notation from Figure 6.3.2, we have the following.

$$\langle 4_1 | S_1 \rangle = A^1 A^{-3} = A^{-2}$$
$$\langle 4_1 | S_2 \rangle = A^4$$

Given a link L with n crossings, the explicit formula for the Kauffman bracket polynomial, $\langle L \rangle$, is obtained by considering all possible states, S, of L. Since each crossing can be smoothed in two distinct ways, L will have 2^n associated states. We take the sum over all states S of L of the product of $\langle L | S \rangle$ with a factor that accounts for the number of connected components in that state.

Theorem 6.3.5. *Given a link L with n crossings, the Kauffman bracket of L is given by*

$$\langle L \rangle = \sum_S \langle L | S \rangle (-A^2 - A^{-2})^{|S|-1}$$

where $|S|$ denotes the number of components of the smoothed state S of L.

In our Figure 6.3.2 example, the summands of $\langle 4_1 \rangle$ corresponding to the states S_1 and S_2 are:

$$\langle 4_1 | S_1 \rangle (-A^2 - A^{-2})^{|S_1|-1} = A^1 A^{-3}(-A^2 - A^{-2}) = -1 - A^{-4}$$
$$\langle 4_1 | S_2 \rangle (-A^2 - A^{-2})^{|S_2|-1} = A^4(-A^2 - A^{-2})^2 = A^8 + 2A^4 + 1$$

Exercise 6.3.6. Show that the Kauffman bracket of the figure eight knot diagram in Figure 6.3.2 is

$$A^8 - A^4 + 1 - A^{-4} + A^{-8}$$

by completing the computation started above and using Theorem 6.3.5. (Hint: There are 14 more states to consider.)

Exercise 6.3.7. Calculate the Kauffman polynomial of the left-handed trefoil knot using the explicit formula for the Kauffman polynomial in Theorem 6.3.5. Compare your answer with the one you obtained in Exercise 6.2.3.

Exercise 6.3.8. Using the intuition you gained from completing Exercises 6.3.6 and 6.3.7, prove Theorem 6.3.5. (Hint: Proceed by induction on the number of crossings in a knot diagram.)

Exercise 6.3.9. Conjecture and prove a relationship between the Kauffman bracket polynomial of a link L and its mirror image L^m. (Hint: First, look at the relationship between the Kauffman bracket of the left- and right-handed trefoils for inspiration. Next, test your conjecture. The knot 4_1 is equivalent to its mirror image. Is your conjecture consistent with this fact?)

Now that we are armed with an explicit formula for the Kauffman bracket, we can work toward a proof of our main result. Our proof of Tait's Conjecture makes use of the Kauffman bracket via its *span*, defined as follows.

Definition 6.3.10. For a link diagram L, the **Kauffman span** of L, span(L), is the maximum degree of the Kauffman bracket of L, $\langle L \rangle$, minus the minimum degree of $\langle L \rangle$.

Exercise 6.3.11. (a) What is span(4_1)? (b) How does the span of the left-handed trefoil compare to the span of the right-handed trefoil?

Interestingly, while the Kaufman bracket fails to be a link invariant, its span is a link invariant.

Theorem 6.3.12. *The Kauffman span of a link L, span(L), is a link invariant.*

Exercise 6.3.13. Prove Theorem 6.3.12. (Hint: Make use of the results of Section 6.2.)

Our strategy for proving Tait's Conjecture will involve determining the span of a connected alternating link diagram.

Exercise 6.3.14. Let c be a crossing of connected link diagram L. Suppose the link diagram L is checkerboard colored such that, at the crossing c, the shaded regions adjacent to c are the regions that would be joined by an A-smoothing at c. Show that if L is alternating, then at *every other crossing in the diagram*, the shaded regions are the regions that would be joined by an A-smoothing of the crossing. (See Figure 6.3.3 for examples of such checkerboard colorings.)

Figure 6.3.3: The left-handed and right-handed trefoils with A-smoothing checkerboard colorings.

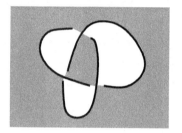

Our next goal is to prove the following lemma about the maximum and minimum degrees of the variable A in the Kauffman bracket.

Lemma 6.3.15. *Suppose L is a connected, reduced, alternating link diagram. Suppose L is checkerboard colored so that at each crossing of L the shaded regions are precisely the regions joined by an A-smoothing of the crossing. Then the terms of highest and lowest degree in $\langle L \rangle$ are given by*

$$\textbf{max degree term} = (-1)^{W-1} A^{cr(L)+2W-2}$$

$$\textbf{min degree term} = (-1)^{B-1} A^{-cr(L)-2B+2},$$

where $cr(L)$ is the number of crossings in L, W is the number of unshaded regions in the diagram, and B is the number of shaded regions in the diagram.

Exercise 6.3.16. Checkerboard color the knot 4_1 and use your calculation from Exercise 6.3.6 to verify that Lemma 6.3.15 holds for this example.

We will work toward a proof of Lemma 6.3.15 by completing the following exercises. **Assume that the hypotheses of Lemma 6.3.15 hold for Exercises 6.3.17, 6.3.18, 6.3.19, and 6.3.20.**

Exercise 6.3.17. Let S be the state of all A-smoothings of the crossings of L. Calculate the contribution of S to the Kauffman bracket $\langle L \rangle$. Show that the max degree term of this contribution is

$$(-1)^{W-1} A^{cr(L)+2W-2}.$$

To prove Lemma 6.3.15, it suffices to show that this max degree term from state S (with all A-smoothings) does not cancel with a term coming from the contribution of some other state of L.

Exercise 6.3.18. Let S be the state of all A-smoothings and let S' be some other state. Observe that S' can be obtained from S by changing one smoothing at a time from an A-smoothing to an A^{-1}-smoothing. Let $S = S_0, S_1, S_2, \ldots, S_k = S'$ be a sequence of states of L that results in S' such that each state has exactly one more A^{-1} smoothing than the previous listed state. Prove that the maximal degree contribution of S_{i+1} is less than or equal to the maximal degree contribution of S_i.

Exercise 6.3.19. Use the fact that the diagram L is *reduced* to prove that the maximal degree contribution of S_1 is strictly less than that of S_0.

Exercise 6.3.20. The previous exercises imply that the maximum degree term of $\langle L \rangle$ is

$$(-1)^{W-1} A^{cr(L)+2W-2}.$$

Follow steps similar to Exercises 6.3.17, 6.3.18, and 6.3.19 to prove that the minimum degree term of $\langle L \rangle$ is

$$(-1)^{B-1} A^{-cr(L)-2B+2}.$$

Exercise 6.3.21. Suppose that L is a connected, reduced, alternating link diagram. Use the previous results to determine the span of L. (Note that the span will be in terms of the number of crossings, $cr(L)$, in L and the numbers of shaded and unshaded regions, B and W, in a certain checkerboard coloring of L.)

Before we complete the proof of Tait's Conjecture, we need to take a brief and helpful detour to learn about a topological invariant called the *Euler characteristic*, discovered around 1750 by the great mathematician Leonhard Euler.

Definition 6.3.22. Suppose X is a topological space composed of V vertices, E edges, and F 'faces' or regions. Then the **Euler characteristic** of X, denoted $\chi(X)$, is defined by the formula

$$\chi(X) = V - E + F.$$

To see how the Euler characteristic is computed, let us look at one of the simplest topological spaces: a sphere. It is not hard to show that the Euler characteristic of the sphere is equal to 2. Indeed, the sphere can be decomposed into one vertex (say, somewhere on the equator), an edge that travels along the equator and is bounded on both ends by the single vertex, and two 'faces' or regions consisting of the northern and southern hemispheres, as in Figure 6.3.23. In this case, $V = 1$, $E = 1$, and $F = 2$, and so $\chi = 1 - 1 + 2 = 2$.

Figure 6.3.23. A sphere composed of one vertex, one edge, and two faces.

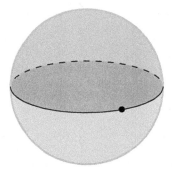

What makes the Euler characteristic so useful is that it is a *topological invariant*. In other words, no matter how we decide to view a topological space X as a collection of vertices, edges, and faces, $\chi(X)$ remains unchanged. In particular, there are any number of other ways we could have chosen to divide the sphere into vertices, edges, and faces, but regardless of which choice we make, $\chi = 2$.

You might wonder, how does this relate to our current situation? Suppose we draw a connected link diagram L on the surface of a sphere. If we forget L's crossing information, we can view the diagram as a graph, where the crossings in L are vertices and the arcs between adjacent crossings are edges. The regions bounded by the edges of the graph form faces on the surface in which L lies. The outer region of a link diagram in the plane forms a strange sort of face. Notice that, if instead of viewing our graph as lying in the plane, we view it as lying on a sphere, this outer face is far more like the other faces in the diagram.

Figure 6.3.24. A sphere decomposed into vertices, edges, and faces by a connected link shadow.

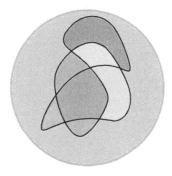

Exercise 6.3.25. Verify that the Euler characteristic of the sphere S is 2, using the decomposition of S into vertices, edges, and faces created by the knot shadow pictured on the sphere in Figure 6.3.24.

Next, in order to harness the Euler characteristic to relate the number of crossings to the number of regions in a link diagram, we need to determine the relationship between the number of crossings and the number of arcs in a link diagram. In other words, what is the relationship between the number of vertices and edges in the graph created by a link shadow?

Exercise 6.3.26. View the shadow of a connected link diagram L as a graph on a sphere by forgetting the crossing information. Conjecture and prove a relationship between the number of vertices and the number of edges in the graph in terms of the number of crossings of L. (Hint: Draw three different link shadows, each with 5–10 precrossings. For each example shadow, count the number of vertices and edges in the graph and compare with the number of crossings in your link diagram.)

Now we are convinced that the shadow of a connected link diagram L gives rise to a connected planar graph on the sphere such that $V = cr(L)$,

$E = 2cr(L)$, and $F = B + W$. Thus,

$$
\begin{aligned}
2 &= \chi(S) \\
&= V - E + F \\
&= cr(L) - 2cr(L) + (B + W).
\end{aligned}
$$

It follows that
$$B + W = 2 + cr(L).$$

At last, we can use this relationship between crossings and regions to prove the main result.

Exercise 6.3.27 (Proof of Tait's Conjecture). Prove that any two connected, reduced, alternating link diagrams for a given alternating link have the same number of crossings. (Hint: Let L and L' be two connected, reduced, and alternating diagrams of the same link. Prove that $cr(L) = cr(L')$ by observing that span(L) = span(L'), and then simplifying by using Exercise 6.3.21 together with the result we just derived.)

Chapter 7

Unknotting Operations & Invariants

7.1   Unknotting Operations

One interesting measure of complexity in knot theory concerns how difficult it is to unknot a nontrivial knot if certain unknotting operations are allowed. In this section, we explore several methods of unknotting. We will then be able to consider some questions related to knot complexity.

As we saw in Section 1.4, the most fundamental unknotting operation is that of a **crossing change**, pictured in Figure 7.1.1 below.

Figure 7.1.1: The crossing change move.

Exercise 7.1.1. Suppose the knot diagrams in Figure 7.1.2 need to be unknotted. Perform a number of crossing changes on these diagrams to produce diagrams of the unknot. In each case, how many crossing changes did you need? (Could you have used fewer?)

Figure 7.1.2: Knot diagrams to be unknotted.

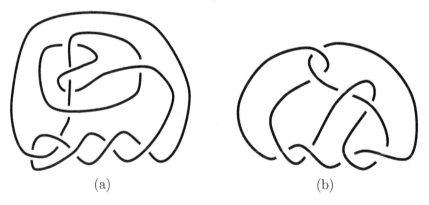

(a)   (b)

Exercise 7.1.2. Using what you discovered in Section 1.4, prove the following. Given any knot diagram, a number of crossing changes can be performed on the diagram to produce a diagram of the unknot.

Exercise 7.1.3. Prove that no more than half the crossings in a knot diagram must be changed in order to produce a diagram of the unknot. (Hint: The mirror image of a diagram of an unknot is also a diagram of the unknot. Prove this and illustrate how this fact relates to the exercise.)

While the crossing change move is the standard unknotting operation, there are several more local moves that can be used to produce the unknot. It may not be obvious to see why, but the #-move and the ▲-move, defined in Figure 7.1.3, are two more examples of unknotting operations in the following sense. Given any knot diagram D, there is a sequence of Reidemeister moves and #-moves (or ▲-moves) that can be performed on D to produce an unknotted circle.

Figure 7.1.3: The unknotting #-move and ▲-move.

These unknotting operations (and others) are studied in [2, 26, 27].

Another operation that can be applied to knot diagrams is called the **region crossing change**, or **RCC move**. Given a diagram of a knot, an RCC move is performed by choosing a region R in the diagram and changing all crossings that fall along the boundary of that region. An illustration of how this move works is shown in Figure 7.1.4. We can also allow the RCC move to be performed on the region surrounding the knot.

Observe that the #-move shown in Figure 7.1.3 is just a special type of RCC move.

Exercise 7.1.4. Consider the knot diagrams in Figure 7.1.2. For each diagram, perform an RCC move on the region *outside* of the diagram. What does the resulting diagram look like? In each case, how many crossings were affected by the RCC move?

Exercise 7.1.5. Consider the knot diagrams (a1) and (b1) on the left in Figure 7.1.4. Perform RCC moves on some number of regions in each diagram to produce diagrams of the unknot. What is the fewest number of moves needed to unknot each diagram?

Figure 7.1.4: The RCC move performed on a region R in two different knot diagrams.

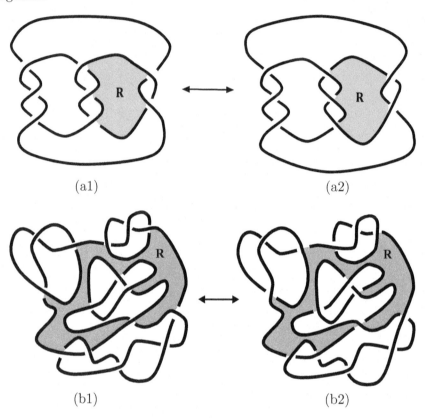

(a1)   (a2)

(b1)   (b2)

Exercise 7.1.6. Determine the effect of performing an RCC move *twice* on the same region in a knot diagram. Formulate and prove a conjecture about the effect of performing the RCC move an even number of times on a given region of a knot diagram.

Exercise 7.1.7. Consider the two knot diagrams in Figure 7.1.2. For each of the two diagrams, find a set of regions such that performing RCC moves on those regions produces an unknot diagram.

Exercise 7.1.8. The knot diagram (a) on the left in Figure 7.1.2 is nonalternating. Find a set of regions on which you can perform RCC moves to produce an alternating knot diagram.

In Exercise 7.1.2, you showed that the crossing change move is an unknotting operation. Perhaps this is an unsurprising fact. It is much more surprising, however, that the RCC move can be used to unknot *any* knot diagram, not just those diagrams you unknotted in Exercises 7.1.5 and 7.1.7. A young Japanese mathematician named Ayaka Shimizu was the first to prove our next main theorem, Theorem 7.1.9, in [37]. Using her strategy, we will prove that the RCC move is an unknotting operation.

Theorem 7.1.9. *Given any knot diagram, it is possible to obtain a diagram of the unknot by performing some number of region crossing change moves on the diagram.*

Our global strategy to prove Theorem 7.1.9 will be to piggyback on the result that the crossing change (CC) move is known to be an unknotting operation, as we showed in Exercise 7.1.2. So it suffices to prove that, for an arbitrary crossing c in a knot diagram D, there exists a collection of regions in D on which RCC moves can be performed to change c and only c. This will prove that the RCC move is also an unknotting operation since each crossing change that is required to unknot the diagram can be achieved by performing RCC moves on some collection of regions in the diagram. Keeping this in mind, let's focus our efforts to prove the following proposition. Theorem 7.1.9 will follow as a corollary.

Proposition 7.1.10. *Given any knot diagram D and a crossing c in D, there exists a collection of regions in D on which RCC moves can be performed in order to change the crossing c and only crossing c in D.*

Our goal is to prove Proposition 7.1.10 in two cases, first for the special case of *reduced knot diagrams*, as in Definition 6.3.1. Then, we will construct an argument by induction on the number of reducible crossings to prove the proposition in general. To see why these two cases may need separate treatment, we make the following observations.

Figure 7.1.5: Regions A, B, C, and D adjacent to a crossing.

Exercise 7.1.11. Consider region A in Figure 7.1.5.

1. Assuming this crossing appears in an arbitrary knot diagram, determine which of the remaining regions (if any) *could* actually be the same region as A. (Hint: It may help to think about this question in terms of the crossings in a reducible diagram, e.g., diagrams (b1) and (b2) in Figure 7.1.4.)

2. Assuming this crossing appears in a reduced knot diagram, determine which of the remaining regions (if any) could be the same region as A.

3. Summarize your observations.

Exercise 7.1.12. Suppose the crossing in Figure 7.1.5 appears in a reduced knot diagram. Determine the effect on the crossing of performing RCC moves on:

(i) an even number of the regions A, B, C, and D;

(ii) an odd number of the regions A, B, C, and D.

Explain why the assumption that our knot diagram is reduced is important.

In Definition 5.5.4, we introduced the idea of a checkerboard coloring of a knot or link diagram. As it happens, checkerboard colorings play an important role in figuring out which RCC moves should be performed to change exactly one crossing in a diagram. To continue setting the stage to prove Proposition 7.1.10 in the case where the diagram D is reduced, let's work through the following exercises.

Exercise 7.1.13. Consider the knot diagram (a1) in Figure 7.1.4, ignoring the shading in the region labeled R. Draw this diagram together with a checkerboard coloring. Determine what happens when RCC moves are performed on *all* of the shaded regions.

Exercise 7.1.14. Suppose, more generally, that you are given a *reduced knot* diagram D and its checkerboard coloring. Formulate and prove a conjecture that describes the effect of performing RCC moves on all of the shaded regions in the checkerboard coloring of D.

With these observations in hand, let's begin to prove Proposition 7.1.10 for the reduced diagram case. The following three-step algorithm can be used to determine the regions on which RCC moves will be applied to achieve our desired crossing change in a reduced diagram.

Step 1: Assign an orientation to the reduced knot diagram D and smooth the diagram at crossing c with respect to the orientation, as pictured in Figure 7.1.6.

Exercise 7.1.15. Explain why the choice of orientation in Step 1 does not influence the resulting smoothed diagram.

Exercise 7.1.16. Explain why the diagram that results from Step 1 is a two component link. How do the two arc segments that come from the smoothing contribute to the two components?

After Step 1, the new diagram will contain a newly formed region which consists of two regions from the original knot diagram. Call this new region R, as pictured in Figure 7.1.6.

Figure 7.1.6: Smoothing at a positive or negative crossing in the direction of the orientation.

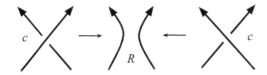

Step 2: After smoothing at crossing c, select one of the two components in the link diagram. For the selected component, *completely ignoring the other*, checkerboard color it so that the region R is unshaded.

To clarify Step 2, Figure 7.1.7 illustrates two examples where one component of a link is checkerboard colored while the other component is ignored.

Step 3: Using the shaded regions from Step 2, shade the corresponding regions in our original knot diagram and perform RCC moves on precisely these regions (exactly once per region).

Exercise 7.1.17. Apply the three-step algorithm as follows to the reduced diagram in Figure 7.1.8 to see how the algorithm works.

Figure 7.1.7: Checkerboard coloring a single link component in a diagram with two components.

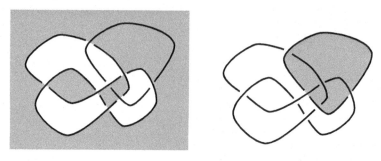

Figure 7.1.8: A knot to be simplified via RCC moves.

1. Apply Step 1 to the diagram in Figure 7.1.8, smoothing at the crossing labeled c. Label the newly formed region R.

2. Choose one component of this diagram to checkerboard color and shade regions following Step 2. Note that one component choice will result in a rather trivial checkerboard coloring (where essentially all that needs to be done is to shade the interior of a disk) while the other component choice leads to a more interesting checkerboard coloring. Either choice will work.

3. Follow Step 3 to identify the regions in the original knot diagram we can apply RCC moves to in order to change crossing c and only crossing c. Verify that performing the RCC move on these regions gives us our desired result.

Now, let's back up and see what would have happened if we had made the other component choice in Step 2.

Exercise 7.1.18. You made a choice in Exercise 7.1.17 about which link component to checkerboard color. Now, go back to your smoothed diagram and select the other component. Follow Steps 2 and 3 to find an alternate collection of RCC moves that will switch crossing c and only crossing c. Verify that your collection of regions changes crossing c and only crossing c.

Exercise 7.1.19. Go through steps 1–3 once again to identify a set of regions on which RCC moves can be applied in Figure 7.1.8 in order to change crossing c' and only c'.

Now, we have two sets of regions: one set of regions on which RCC moves can be performed to change c, and another set of regions on which RCC moves can be performed to change c'. What if we want to change *both* crossings? In the next exercise, we'll explore how to harness the previous results to perform more than one crossing change at a time.

Exercise 7.1.20. Identify a set Q of regions of the diagram D in Figure 7.1.8 for which the following holds. If an RCC move is applied *exactly once* to each region in Q, both crossings c and c' will change, and all other crossings will stay the same. (Hint: Use your result from Exercise 7.1.12.)

The examples above give good evidence that the three-step algorithm indeed determines a set of regions in a reduced diagram for which the application of RCC moves will successfully change a specified crossing. Now we must *prove* that this process works in general.

Lemma 7.1.21 (Proposition 7.1.10: The Reduced Diagram Case). *Given a reduced diagram, D, and a crossing c in D, there exists a collection of regions in D on which RCC moves can be performed in order to change the crossing c while leaving all other crossings in D unchanged.*

Exercise 7.1.22. Prove Lemma 7.1.21. In other words, show that the three-step algorithm always works. (Hint: There are two natural steps in this proof. (1) Prove that if RCC moves are performed on the identified collection of regions, then crossing c changes. (2) Prove that, for a crossing $d \neq c$ in the diagram, d remains unchanged after the RCC moves have been performed. Be sure you know why your proof only works for reduced knot diagrams!)

Now that we have shown Proposition 7.1.10 holds for reduced knot diagrams, let's consider diagrams with reducible crossings. We will work toward understanding which regions to select in an arbitrary reducible diagram by first considering the case of a knot diagram with *exactly one* reducible crossing. A careful analysis of this case will form the foundation for an inductive argument on the number of reducible crossings in a knot diagram. Our next goal, then, will be to prove the following lemma.

Lemma 7.1.23 (Proposition 7.1.10: The Case of Exactly One Reducible Crossing). *Given a diagram, D, with exactly one reducible crossing d, and a specified crossing c (possibly equal to d) in D, there exists a collection of regions in D on which RCC moves can be performed in order to change the crossing c while leaving all other crossings in D unchanged.*

To prove Lemma 7.1.23, we consider two cases. In Case 1, the crossing d is precisely the crossing we wish to change. In Case 2, we would like to change a reduced crossing, $c \neq d$, of D. Let's explore an example to determine how to attack the simpler case, Case 1.

Exercise 7.1.24. The diagram in Figure 7.1.9 has one reducible crossing d. Determine how *a portion of the diagram* can be checkerboard colored such that, if an RCC move is performed on each shaded region, the reducible crossing d and only d is changed.

Figure 7.1.9: A reducible diagram with exactly one reducible crossing, d.

Using this example for intuition, prove Case 1 of Lemma 7.1.23.

Exercise 7.1.25. (Proof of Case 1 of Lemma 7.1.23) Describe, in general, how to identify a collection of regions in a diagram, D, with exactly one reducible crossing, d, such that using RCC moves on these regions changes the crossing d and only d. Prove that the regions described achieve the desired crossing change. (Hint: You may wish to use the notation or smoothing idea described in Figure 7.1.10.)

Next, we turn to Case 2 of Lemma 7.1.23 where we want to change a crossing c that is a reduced crossing in a reducible diagram D. For this case, the following notation will come in handy. By definition, the reducible crossing d has an associated region that is twice adjacent to d; call this associated region A. If the diagram D is smoothed at the reducible crossing d with respect to a choice of orientation (as in Step 1 of the three-step algorithm), then the smoothed diagram is a disjoint union of two diagrams with the region A separating them. Of these two diagrams, we call the component that contains c diagram D_2. We'll call the other reduced diagram D_1. Furthermore, in the diagram D, let B denote the region adjacent to crossing d such that after smoothing at d, B can be viewed as a region inside D_2. See Figure 7.1.10 for an illustration of these naming conventions.

Figure 7.1.10: A diagram with a reducible crossing, d. Before smoothing, the regions A and B are adjacent to the crossing d, and region A is twice adjacent to d. After smoothing at d, D_1 is the diagram that does not contain crossing c.

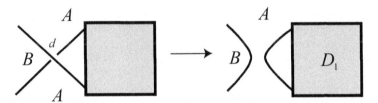

Notice that in Case 2 of Lemma 7.1.23, D contains exactly one reducible crossing, d. If we smooth at d, there are no more reducible crossings in the resulting link diagram. In other words, both of the diagrams D_1 and D_2 are reduced knot diagrams. By Lemma 7.1.21, the three-step algorithm can be used on D_2 (while ignoring D_1) to determine a collection of regions of D_2 that change c and only c. A reasonable guess is that this collection of regions, if changed in the original diagram D, will give us our desired result. However, there are some subtleties in how these selected regions behave when viewed as regions of D. Let's look at an example to give us ideas for a general proof of Case 2.

Figure 7.1.11: The diagram from Figure 7.1.9 with the reducible crossing d smoothed and the knot diagram D_1 ignored. In D_2, the crossing c has been smoothed and one component of the resulting link is checkerboard colored (so that R is unshaded) while the other link component is ignored. Notice that the region A_2 of D_2 is the entire large region at the center of the diagram containing D_1, since D_1 is ignored.

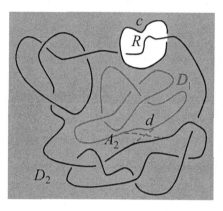

Exercise 7.1.26. Return to the example in Figure 7.1.9. Smooth the reducible crossing d, respecting the orientation of the knot. Follow the three-step algorithm on D_2 (completely ignoring D_1) and *determine the selected regions of D_2* that will change c and only c in D_2. Regardless of the choice of component you made in the three-step algorithm, the region A should be shaded. One such component choice is shown in Figure 7.1.11. Draw the checkerboard coloring corresponding to the *other choice*.

We now take the regions of D_2 that were selected in Exercise 7.1.26 and adjust them to find regions in the original diagram D on which we should perform RCC moves. You might ask, why do the regions need adjusting? To answer this question, let's view all of the shaded regions in Figure 7.1.11 as regions of D and pay special attention to what happens at d if RCC moves are applied to all of these regions. See Figure 7.1.12.

The crossing c will indeed be changed by these regions, since RCC moves are applied to exactly one of the four regions adjacent to c. Unfortunately, the crossing d is *also* changed when RCC moves are applied to all of the shaded regions in Figure 7.1.12, since d only has three distinct adjacent regions, and all of them are shaded. So, in this example, a modification of the selected regions is required to determine an unknotting set of regions.

Figure 7.1.12: When RCC moves are performed in the shaded regions of D, *two* crossings change in the diagram.

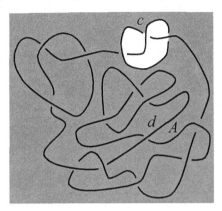

It should not be surprising that the adjustments we'd like to make are in the portion of the diagram that has thus far been ignored, namely D_1.

Exercise 7.1.27. Modify the shaded regions in the diagram in Figure 7.1.12 by *unshading* certain regions so that, if we perform RCC moves on all of the remaining shaded regions, c and only c is changed.

Exercise 7.1.28. Use the checkerboard coloring corresponding to the *other component choice* that you found in Exercise 7.1.26 to determine whether performing RCC moves on all of the shaded regions of your diagram (viewed as regions of D) changes c and only c in D. Is a modification to your collection of regions needed? Why or why not? What adjustment to the set of regions (if any) will allow us to change c and only c by performing RCC moves on the regions in the set?

Using Exercises 7.1.27 and 7.1.28 for intuition, let's go back to the general setting of Case 2 of Lemma 7.1.23. We will let $\mathcal{Q}$ denote our desired collection of regions of D on which RCC moves can be applied to change crossing c and only c. By Lemma 7.1.21, since D_2 is reduced, there exists a collection of regions of D_2, call it $\mathcal{P}$, such that performing RCC moves on these regions achieves the desired crossing change at c and only c in D_2. (Once again, we are ignoring D_1 and its regions, as in Figure 7.1.11.) Let's call A_2 the region of D_2 that contains A, as illustrated in two different types of examples in Figure 7.1.13. Since the regions A_2 and B may or may not be in $\mathcal{P}$, there are four mutually exclusive possibilities to consider in order to determine what $\mathcal{Q}$ should be. We begin by including in $\mathcal{Q}$ all of

Figure 7.1.13: Two examples illustrating the relationship between A in D and A_2 in D_2.

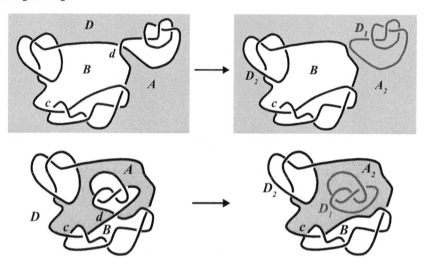

the regions in $\mathcal{P}$—each now viewed as a region of D—*except for* A_2 if A_2 is in $\mathcal{P}$ (since A_2 is not a region of D).

Here are the four subcases.

1. Suppose $\mathcal{P}$ contains both regions A_2 and B, as in Figure 7.1.14. Checkerboard color D_1 (which sits inside A_2) in such a way that the surrounding region A is included in the set of shaded regions. Add to $\mathcal{Q}$ both A and all the newly shaded regions of D_1 that lie within A_2.

   Figure 7.1.14: Subcase 1.

   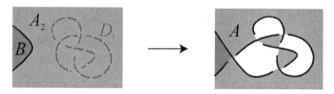

2. Suppose A_2 is in $\mathcal{P}$ and B is not in $\mathcal{P}$, as in Figure 7.1.15. In this case, add to $\mathcal{Q}$ all regions inside D_1, as well as A.

3. Suppose A_2 is not in $\mathcal{P}$ and B is in $\mathcal{P}$. Determining the set of regions, $\mathcal{Q}$, in this case is left as Exercise 7.1.29.

Figure 7.1.15: Subcase 2.

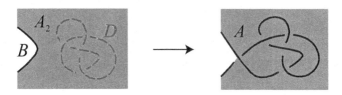

4. Finally, suppose that A_2 is not in $\mathcal{P}$ and B is not in $\mathcal{P}$, as in Figure 7.1.16. Then neither regions inside D_1 nor region A should be included in Q.

Figure 7.1.16: Subcase 4.

Exercise 7.1.29. Complete the description of regions $\mathcal{Q}$ in Subcase 3, assuming A_2 is not in $\mathcal{P}$ and B is in $\mathcal{P}$. Provide a figure in support of your description, analogous to the figure for each case given above.

Exercise 7.1.30. (Proof of Case 2 of Lemma 7.1.23) For each of the four subcases above, prove that if $\mathcal{Q}$ contains exactly the regions specified, then performing RCC moves on all regions in $\mathcal{Q}$ will change c in D and no other crossings.

We are now more than ready to prove Proposition 7.1.10.

Proposition 7.1.31 (Inductive Step of Proposition 7.1.10)**.** *Suppose D is a reducible knot diagram containing at least one reducible crossing. Let c be any crossing in D, perhaps a reducible crossing. Then there is a subset of regions Q of D on which RCC moves can be performed in order to change crossing c, while leaving all other crossings in D unchanged.*

Proof of Proposition 7.1.31. We will argue by induction on k, the number of reducible crossings in D. Lemma 7.1.21 states that Proposition 7.1.10 holds for a reduced diagram, that is $k = 0$, while

Lemma 7.1.23 states that Proposition 7.1.10 holds for $k = 1$. This 'second' base case holds the key to the inductive step! At the heart of the proof of Lemma 7.1.23 (Case 2), we smoothed at a reducible crossing, leaving two disjoint diagrams, D_1 and D_2, where we assumed that D_2 contains c and one fewer reducible crossing.

Suppose Proposition 7.1.31 holds for all knot diagrams with k or fewer reducible crossings. Let c be a crossing in a knot diagram with $k + 1$ reducible crossings.

Exercise 7.1.32. Complete the proof of Proposition 7.1.31. (Hint: Consider two cases, one where c is a reducible crossing and one where c is reduced. First, consider the case where c is reducible, and use what you learned in Exercise 7.1.25. For the second case, argue that there must exist a reducible crossing d in D such that $d \neq c$ and smoothing at d results in two disjoint diagrams: D_2 containing c, and D_1 which is reduced. Then apply the inductive hypothesis to D_2 and complete the proof using four cases analogous to Lemma 7.1.23.)

7.2   The Unknotting Number

In Section 1.2, the **unknotting number** of a knot is defined to be the minimum number of times the knot must pass through itself before it becomes unknotted. In other words, the unknotting number of a knot K is the minimum number of crossing change moves that must be performed on a diagram of K to produce the unknot, where the minimum is taken over all diagrams of K. We denote the unknotting number of K by $u(K)$.

Exercise 7.2.1. Explain why the unknotting number of a knot is a knot invariant.

Exercise 7.2.2. Prove that a knot K has unknotting number 0 if and only if K is the unknot.

Exercise 7.2.3. Prove that all twist knots have unknotting number one.

Exercise 7.2.4. Show that for the $(2, p)$ torus knot, $T_{2,p}$, the unknotting number satisfies $u(T_{2,p}) \leq \frac{p-1}{2}$. (In fact, $u(T_{2,p}) = \frac{p-1}{2}$, but equality is more difficult to prove! Note for this problem that p must be odd for $T_{2,p}$ to be a knot rather than a link.)

For cases where it is less obvious how to unknot a knot diagram, we can use the following easily computable upper bound to determine partial information about the unknotting number. Suppose we have a knot K with diagram D. Form the Gauss diagram G_D associated to D, as described in Chapter 4. The **trivializing number**, $t(D)$, is equal to the minimum number of arrows that must be removed from G_D so that no arrows intersect.

For instance, in Figure 7.2.1, we picture a twist knot, its Gauss diagram, and a subdiagram of the Gauss diagram containing only nonintersecting arrows. In this example, $t(D) = 2$ since two arrows (1 and 5) had to be removed to produce a diagram with only nonintersecting arrows.

Figure 7.2.1: A knot diagram D, its Gauss diagram G_D, and a subdiagram of G_D containing no intersecting arrows.

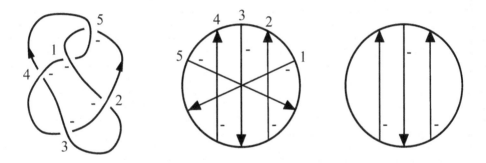

The trivializing number was originally defined by Ryo Hanaki in [13]. Building on Hanaki's work, a group of undergraduate researchers working in the Williams College SMALL REU proved the following result [15].

Theorem 7.2.5. *For any diagram D of a knot K,*

$$u(K) \leq \frac{t(D)}{2}.$$

Exercise 7.2.6. Use Theorem 7.2.5 to give an alternative proof that the unknotting number of any twist knot is 1.

Exercise 7.2.7. Consider the knot diagrams in Figure 7.2.2. Find an upper bound on the unknotting number for each knot. (Hint: Find the trivializing number for each diagram and use Theorem 7.2.5.)

Figure 7.2.2: Diagrams to be unknotted.

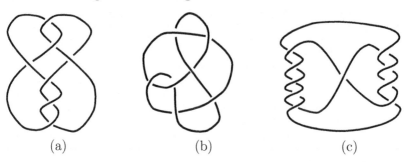

(a)   (b)   (c)

In Exercise 7.2.7, you found upper bounds for the unknotting numbers of several knot diagrams. In particular, you discovered indirectly that diagram (c) can be unknotted with three crossing changes. Surprisingly, if we make the diagram of this knot *more complex*, the knot can be unknotted with *fewer* crossing changes!

Exercise 7.2.8. Observe two diagrams of the pretzel knot $P_{5,1,4}$ in Figure 7.2.3. Verify that (i) the minimal crossing diagram of $P_{5,1,4}$ can be unknotted by changing its three highlighted crossings; and (ii) the diagram with more crossings can be unknotted by changing its two highlighted crossings.

Figure 7.2.3: Nakanishi-Bleiler example.

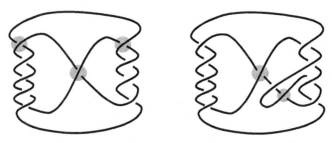

The pair of diagrams in Figure 7.2.3 was discovered independently by Nakanishi [28] and Bleiler [6] as an example showing that the unknotting number of a knot might only be achievable on a diagram that has more than the minimum number of crossings. This is one of the factors that makes the unknotting number of a knot particularly difficult to compute.

One famous result that has been used to determine certain unknotting numbers was proven by Martin Scharlemann [35]. Recall, from Section 1.6, the definitions of prime and composite knots. His classical theorem concerning unknotting numbers that is easy to understand, but difficult to prove, is the following.

Theorem 7.2.9. *Any composite knot has unknotting number at least 2. Equivalently, all knots with unknotting number 1 are prime.*

Exercise 7.2.10. Use Theorem 7.2.9 to answer the following questions.

1. What can you conclude about the unknotting number of the square knot (pictured in Figure 5.5.1)?

2. What does Exercise 7.2.3 imply about the family of twist knots?

7.3 The Region Unknotting Number

We have just explored the classical unknotting number. Similarly, we can define unknotting numbers for the other unknotting operations, just as Ayaka Shimizu first did for the RCC move in [37]. In particular, the **region unknotting number**, $u_R(K)$, of a knot K is the minimum number of RCC moves that must be performed in a diagram of K to produce the unknot, where the minimum is taken over all diagrams of K. Abusing notation, we use $u_R(D)$ to denote the minimum number of regions that must be changed in a particular diagram D of a knot in order to produce a diagram of the unknot. Note that if D is a diagram of K, then $u_R(K) \leq u_R(D)$. (Explain why!)

Exercise 7.3.1. Prove that the following families of knots all have region unknotting number 1.

1. Twist knots

2. Knots with Conway notation $[[m\ 2\ m]]$ (as in Definition 4.3.40)

3. Knots with Conway notation $[[m\ 2\ (m \pm 1)]]$

Exercise 7.3.2. Make conjectures about the values of $u_R(K)$ for each of the knots K shown in Figure 7.2.2.

Exercise 7.3.3. Make a conjecture about the region unknotting number of a $(2, p)$ torus knot, $u_R(T_{2,p})$.

In an attempt to learn something about the region unknotting number, one natural question to ask is: what is the relationship between $u_R(K)$ and the crossing number $c(K)$ of a knot K? A related, but simpler, question to answer is the following: what is the relationship between the region unknotting number of a knot diagram and the number of regions in the diagram? Let's explore these questions for the case of reduced diagrams of knots.

The following terminology will help us in our exploration. Suppose we have a reduced knot diagram with a checkerboard coloring. Let B be the set of shaded regions in the diagram and W be the set of unshaded regions. Now suppose that P is some subset of B and V is some subset of W. Then, $B - P$ will denote all shaded regions *not* in the subset P and $W - V$ will denote all unshaded regions *not* in V. Let's proceed by working through the following two exercises, using what we learned in Exercise 7.1.14 about applying RCC moves to all shaded regions in a reduced knot diagram.

Exercise 7.3.4. An example of a reduced knot diagram with sets of regions B, P, W, and V is shown in Figure 7.3.1.

1. Consider the effect of performing RCC moves on all regions in P.

2. Consider the effect of performing RCC moves on all regions in $B - P$.

3. Consider the effect of performing RCC moves on all regions in V.

4. Consider the effect of performing RCC moves on all regions in $W - V$.

What did you notice?

Exercise 7.3.5. Formulate and prove a conjecture about the relationship between the diagram obtained by performing RCC moves on all regions in P and the diagram obtained by performing RCC moves on all regions in $B - P$. Then, formulate and prove a similar conjecture for V and $W - V$.

The following exercise uses the **floor function**, $\lfloor x \rfloor$, i.e., the function that takes in a real number x and returns the largest integer less than or equal to x.

Exercise 7.3.6. Suppose that D is a reduced, checkerboard colored knot diagram and let b denote the number of shaded regions and w the number

Figure 7.3.1: Subsets of checkerboard colored regions: B, the set of 'shaded' regions, consists of the regions colored dark and light green, and W, the set of 'unshaded' regions, consists of regions colored dark and light orange. In particular, regions in the set P are colored dark green, and regions in $B - P$ are light green, while regions in the set V are colored dark orange, and regions in $W - V$ are light orange.

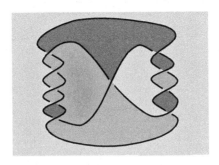

of unshaded regions in D. Let U be a diagram of the unknot with the same underlying projection as D. Show that U can be obtained from D using no more than n moves, where

$$n = \left\lfloor \frac{b}{2} \right\rfloor + \left\lfloor \frac{w}{2} \right\rfloor.$$

(Hint: Use the result of Exercise 7.3.5.)

Exercise 7.3.7. Building upon Exercise 7.3.6, prove Theorem 7.3.8.

Theorem 7.3.8. *The region unknotting number $u_R(D)$ of a reduced knot diagram D is no greater than half the number of regions in D.*

This is a nice result in and of itself, but let's push our investigation a bit further to answer our original question. How does the region unknotting number of a knot K relate to the *crossing number* of K? Since Theorem 7.3.8 tells us about the relationship between u_R and the number of *regions* in a diagram, we're halfway there.

Recall the relationship we discovered in Section 6.3 between the number of regions, F, in a knot diagram D and the number of crossings, $c(D)$:

$$F = 2 + c(D).$$

Exercise 7.3.9. Prove Theorem 7.3.10. (Hint: Use the fact that $F = 2 + c(D)$ together with your result from Exercise 7.3.6.)

Theorem 7.3.10. *Let D be a reduced knot diagram and $c(D)$ the number of crossings in D. Then*

$$u_R(D) \leq \frac{c(D) + 2}{2}.$$

Now, we have all the tools we need to show the following corollary about the region unknotting number of a *knot*, not just the region unknotting number of a particular *diagram* of the knot.

Corollary 7.3.11. *Let K be a knot.*

1. *If D is any reduced diagram of K, then $u_R(K) \leq \frac{c(D)+2}{2}$, where $c(D)$ denotes the number of crossings in D.*

2. *In general, $u_R(K) \leq \frac{c(K)+2}{2}$, where $c(K)$ denotes the crossing number of K, as defined in Activity 1.4. (Hint: Use the fact that every knot has a reduced diagram.)*

Exercise 7.3.12. Prove Corollary 7.3.11.

Chapter 8

Virtual Knots

8.1   What is a Virtual Knot?

Think back to Chapter 4 where we first learned about using Gauss diagrams to record knot diagrammatic information. Since there is a well-defined algorithm for creating a Gauss diagram from a knot diagram, we can see that every knot diagram corresponds to a Gauss diagram. You might ask, though, about the converse of this statement. Does every well-formed Gauss diagram correspond to a knot diagram? The answer is a resounding no. The surprisingly simple example in Figure 8.1.1 illustrates this fact.

Figure 8.1.1: A Gauss diagram without a knot.

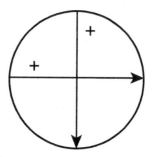

Exercise 8.1.1. Consider the Gauss diagram from Figure 8.1.1. *It is impossible to construct a corresponding knot diagram.* Try to do the impossible and construct a knot diagram for this Gauss diagram. In the process, discover what goes awry. Describe your findings.

This fact is precisely what prompted the discovery of a more general collection of knots called **virtual knots**. A virtual knot can formally be defined as a Gauss diagram, where two virtual knots are considered to be equivalent if and only if their Gauss diagrams can be related by a sequence of the Gauss diagrammatic Reidemeister moves that you discovered in Chapter 4. (See Figure 4.2.5.)

The theory of virtual knots was independently discovered by Louis Kauffman, in [21], and Naoko and Seiichi Kamada, in [17], where the knotlike objects are called **abstract knots**.

Exercise 8.1.2. Show, using the Gauss diagrammatic Reidemeister moves discovered in Chapter 4, that the virtual knots in Figures 8.1.1 and 8.1.2

are equivalent.

Figure 8.1.2: A virtual knot equivalent to the simplest nontrivial virtual knot.

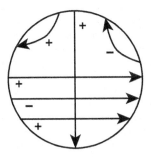

It's clear from the definition that all ordinary knots (i.e., the knots we've been studying until now, which we will henceforth refer to as "classical knots") are also virtual knots since they correspond to Gauss diagrams, but there is a vast world of nonclassical virtual knots. The simplest example is our friend the **virtual trefoil** shown in Figures 8.1.1 and 8.1.2. Figure 8.1.3 illustrates several other simple examples of nonclassical virtual knots.

Figure 8.1.3: Two distinct nonclassical virtual knots.

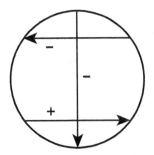 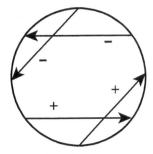

While virtual knots can be defined in terms of Gauss diagrams, they can also be defined in terms of strange sorts of knot diagrams. A virtual knot diagram is a knot diagram that may involve both ordinary crossings and **virtual crossings**. The ordinary, classical crossings are the familiar crossings that appear as arrows in the Gauss diagram. The virtual crossings are those we are forced to draw when reconstructing a knot diagram from a Gauss diagram. Virtual crossings can be thought of as not really existing, since they don't appear in the Gauss diagram of a virtual knot. In a virtual knot diagram, virtual crossings are drawn as in Figure 8.1.4.

Figure 8.1.4: The two types of crossings in a virtual knot diagram: a classical crossing and a virtual crossing.

In Figure 8.1.5, a virtual knot is given both in terms of its virtual knot diagram and its Gauss diagram. For a simple exercise, identify each arrow in the Gauss diagram with its corresponding classical crossing. It is very important to remember that the virtual crossings do not appear in the Gauss diagram—only classical crossing information is recorded.

Figure 8.1.5: A virtual knot diagram and its Gauss diagram.

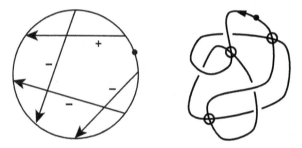

Exercise 8.1.3. Construct virtual knot diagrams for the Gauss diagrams in Figure 8.1.3.

Exercise 8.1.4. Prove that any virtual knot diagram containing *only* virtual crossings (and no classical crossings) must be the unknot. (See Figure 8.1.6 for examples of such diagrams.)

Exercise 8.1.5. Prove that any virtual knot diagram containing *exactly one* classical crossing must be the unknot. (Hint: The proof of this fact may be quite similar to the proof of Exercise 8.1.4.)

Exercise 8.1.6. Construct virtual knot diagrams that correspond to the Gauss diagrams in Figures 8.1.1 and 8.1.2. Can one (or both!) of these

Figure 8.1.6: Examples of virtual diagrams of the unknot.

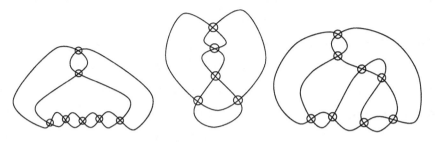

diagrams be drawn with a single virtual crossing? (Hint: Why might this knot be called the virtual trefoil?)

By virtue of our original definition of virtual knot equivalence, we see that classical Reidemeister moves can be performed on virtual knot diagrams without changing the virtual knot type. What other Reidemeister-type moves should be included in a list of R-moves for virtual knots?

Exercise 8.1.7. Any additional Reidemeister moves we allow for knots with virtual crossings should have *no effect on the Gauss diagram* of the knot. In Figure 8.1.7, there are eight *potential* Reidemister-type moves for virtual knots.

1. Determine which subset of these moves we should call the **virtual Reidemeister moves**. Use your results to write a reference sheet of the three classical and four virtual Reidemeister moves that characterize virtual knot equivalence. (Hint: Exactly half of these moves should be allowed.)

2. For each of the moves that should *not* be allowed on our list, provide a Gauss diagram schema illustrating the effect of the prohibited move.

Exercise 8.1.8. Use the set of three classical and four virtual R-moves to show that the virtual knot diagrams you constructed in Exercise 8.1.6 represent equivalent virtual knots. (Notice that this is an alternative way to prove the equivalence in Exercise 8.1.2.)

Exercise 8.1.9. By providing sequences of virtual Reidemeister moves, prove that the diagrams in Figure 8.1.6 are all diagrams of the unknot.

Figure 8.1.7: Eight *potential* virtual Reidemeister moves.

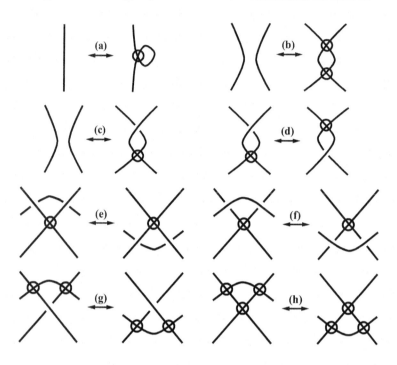

Exercise 8.1.10. Create a virtual knot diagram with eight virtual crossings and six classical crossings that is equivalent to the virtual knot shown in Figure 8.1.5. (Be creative!) Make sure you provide the sequence of R-moves that demonstrates the equivalence.

8.2   Virtual Knot Invariants

Now that we know what virtual knots are, one big question we might ask is the following. How can we tell when a given virtual knot is actually *nonclassical*? In other words, given a Gauss diagram how do we know if we actually need virtual crossings to construct the given knot? Similarly, if presented with a knot diagram that has virtual crossings, how do we know whether it is equivalent to a diagram with only classical crossings? Before we discover a partial answer to this question, let's build some intuition about why this task may be tricky!

Exercise 8.2.1. The diagram in Figure 8.2.1 is a virtual-looking diagram of the unknot. Use the classical and virtual R-moves to prove that this is indeed the unknot.

Exercise 8.2.2. Draw several different virtual knot diagrams that are all equivalent to the (classical) trefoil but that all have some extraneous virtual crossings in them. Show that your diagrams are equivalent to the trefoil by providing a sequence of diagrams that are related by virtual and classical R-moves.

Figure 8.2.1: The unknot in a virtual disguise.

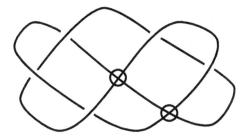

Given a diagram with virtual crossings, it can be difficult to determine whether or not the diagram might represent a classical knot. To identify certain virtual knots that are nonclassical, we introduce an idea of Kauffman's called *crossing parity*, first introduced in [19].

Definition 8.2.3. A crossing in a virtual knot diagram is called **even** if its corresponding chord in the Gauss diagram intersects an even number of chords. Otherwise, we call the crossing **odd**.

Exercise 8.2.4. Classify all chords in the Gauss diagram in Figure 8.1.5 as either even or odd. Use this classification to determine which crossings in the corresponding knot diagram are even and which are odd.

Exercise 8.2.5. Give an alternate definition for even and odd crossings that relies only on the virtual knot diagram, without making reference to the corresponding Gauss diagram. Is it possible to avoid any reference to virtual crossings in your alternate definition?

Now consider the following quantity associated to a knot diagram D.

Definition 8.2.6. The **odd writhe**, $J(D)$, of a knot diagram D is the sum of the signs of all *odd* crossings in D.

Notice that if D has no odd crossings, $J(D) = 0$.

Exercise 8.2.7. Find the odd writhes of the Gauss diagrams in Figures 8.1.1, 8.1.2, 8.1.3, and 8.1.5.

Exercise 8.2.8. Give a virtual knot diagram D for which the value of $J(D)$ is -5.

Exercise 8.2.9. Suppose D is a virtual knot diagram and D^m is its mirror image (obtained from D by changing all classical crossings of D). Show that $J(D) = -J(D^m)$.

In Chapter 5, we learned that the writhe of a knot is not a knot invariant. Interestingly, the odd writhe *is* an invariant for virtual knots.

Exercise 8.2.10. Prove that the odd writhe is an invariant of virtual knots. In other words, prove that J is a function such that if virtual knot diagrams D and D' are equivalent, then $J(D) = J(D')$.

Exercise 8.2.11. Use the result you proved in Exercise 8.2.10 to find a virtual knot diagram D that is not equivalent to its mirror image D^m.

A particularly useful fact about the odd writhe is that $J = 0$ for all classical knots. In other words, if you are able to prove that $J(K) \neq 0$, this is a proof that K is a nonclassical virtual knot!

Exercise 8.2.12. Use the odd writhe to show that one of the Gauss diagrams in Figure 8.2.2 corresponds to a nonclassical knot. Is the other knot classical or not?

Figure 8.2.2: Two virtual knots, given by their Gauss diagrams.

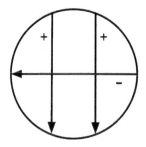 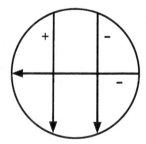

We have seen that the odd writhe is a useful invariant, but there are limits to what it can achieve. Indeed, as you may have discovered, there are nonclassical virtual knots for which $J(K) = 0$. The invariant J only gives us partial information about which virtual knots are classical and which aren't. So what are some other virtual knot invariants? Here, we mention one called the *intersection index polynomial* [7, 14]. To define the intersection index polynomial, we first need to define the intersection index of a single crossing in a virtual knot diagram. Then, we will use the intersection indices of all of the crossings in a virtual knot diagram to form a polynomial invariant for that virtual knot.

For a given oriented virtual knot diagram D, let us choose a classical crossing, d, in this diagram and *smooth* the crossing, as pictured in Figure 8.2.3.

Figure 8.2.3: Smoothing of a crossing with numbered components.

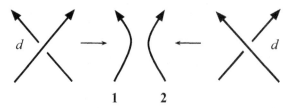

This smoothing produces the diagram of a two-component virtual link. We label the components of the link as follows: when the crossing that is being smoothed is oriented upward, the component on the left is assigned the number 1 and the component on the right is assigned the number 2. Just as in the definition of the linking number of a classical link, we restrict our attention to crossings involving both components. So let C_d denote the set of (classical) crossings in this virtual link diagram that involve both components of the link. We assign signs $\alpha(x)$ to each classical crossing x in C_d as shown in Figure 8.2.4 based on which component of the link each strand of the crossing belongs to. (Note that this is a different way of assigning signs to crossings than we are used to!)

Now we have the following definition. The **intersection index of a crossing** d in a virtual knot diagram K is the sum of the values $\alpha(x)$ for all classical crossings x in C_d. Let's denote the intersection index of crossing d by $i(d)$. See Figure 8.2.5 for an example illustrating how to compute the intersection index for each of the crossings in a given virtual knot diagram.

Exercise 8.2.13. Find the intersection index of each classical crossing in the virtual knot diagram pictured in Figure 8.1.5.

Figure 8.2.4: Definition of $\alpha(x)$ of a classical crossing x in C_d.

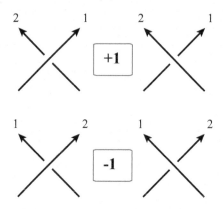

Exercise 8.2.14. Find the intersection index of each classical crossing in the virtual knot diagram pictured in Figure 8.2.6.

Using the idea of an intersection index, we define a polynomial invariant for virtual knots as follows.

Definition 8.2.15. Let the **intersection index polynomial**, $\mathbf{p}_t(K)$, for virtual knot K with diagram D be the sum over all classical crossings d in D of the polynomial $sign(d)(t^{|i(d)|} - 1)$. In more succinct notation:

$$\mathbf{p}_t(K) = \sum_d sign(d)(t^{|i(d)|} - 1).$$

In this formula, the quantity $sign(d)$ is the usual sign of the crossing d (as in Definition 3.4.6), and the letter t is a variable.

For the example in Figure 8.2.5, verify the following computation.

$$\mathbf{p}_t(K) = sign(a)(t^{|i(a)|} - 1) + sign(b)(t^{|i(b)|} - 1)$$
$$+ sign(c)(t^{|i(c)|} - 1) + sign(d)(t^{|i(d)|} - 1)$$
$$= (1)(t^3 - 1) + (1)(t^{|-1|} - 1) + (1)(t^{|-1|} - 1) + (1)(t^{|-1|} - 1)$$
$$= t^3 + 3t - 4$$

Exercise 8.2.16. Show that $\mathbf{p}_t(T_v) = 2t - 2$ for the virtual trefoil T_v by computing the intersection index polynomial in two different ways, using the two equivalent, but distinct, virtual trefoil diagrams you created in Exercise 8.1.6.

Figure 8.2.5: Computing the intersection index for all crossings in a virtual knot diagram.

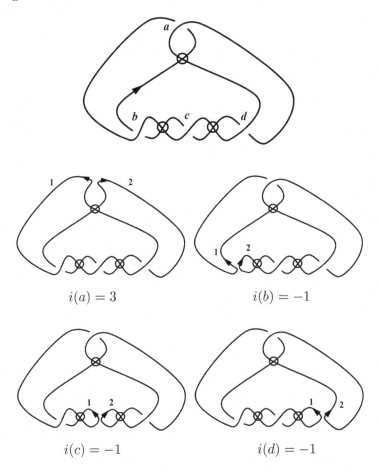

$$i(a) = 3 \qquad\qquad i(b) = -1$$

$$i(c) = -1 \qquad\qquad i(d) = -1$$

Exercise 8.2.17. Find $\mathbf{p}_t(K)$ for the virtual knot K shown in Figure 8.1.5.

Exercise 8.2.18. Find $\mathbf{p}_t(K)$ for the virtual knot K shown in Figure 8.2.6. What is $J(K)$ for this virtual knot?

Exercise 8.2.19. Prove that $\mathbf{p}_t(K)$ does not depend on the virtual knot diagram you choose to represent K. In other words, $\mathbf{p}_t(K)$ is a virtual knot invariant. (Hint: Proceed with your proof by determining the effect of each classical and virtual R-move on the value of the polynomial. In particular, explain why the -1 term is essential for invariance of $\mathbf{p}_t(K)$.)

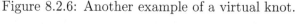

Figure 8.2.6: Another example of a virtual knot.

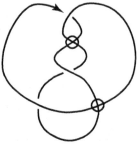

The intersection index polynomial has several interesting properties, as we see in Theorem 8.2.20.

Theorem 8.2.20.   *1. The invariant $\mathbf{p}_t(K) = 0$ if K is a classical knot.*

  *2. The invariant $\mathbf{p}_t(K)$ is **strictly stronger** than J, meaning that if J can be used to prove that two virtual knots K and K' aren't equivalent (by showing $J(K) \neq J(K')$), so can $\mathbf{p}_t(K)$ (by showing $\mathbf{p}_t(K) \neq \mathbf{p}_t(K')$). Furthermore, there are nonequivalent virtual knots that $\mathbf{p}_t(K)$ can distinguish that J can't. (See Exercise 8.2.18.)*

The first statement in Theorem 8.2.20 can be proven by using the famous *Jordan Curve Theorem* to show that the intersection index of any crossing in a classical knot diagram is 0, but let's focus on the second statement. Can you prove it?

Exercise 8.2.21. Prove statement 2 in Theorem 8.2.20. (Hint: Consider the terms corresponding to odd crossings in $\mathbf{p}_t(K)$.)

In this section, we have defined several virtual knot invariants that give us information about the "virtualness" of nonclassical virtual knots, but that are uninteresting for classical knots. There are many more virtual knot invariants that have been studied that not only distinguish nonclassical virtuals, but that can also distinguish classical knots from one another. For instance, there are several ways to extend the Jones polynomial so that it is an invariant of virtual knots but it agrees with the Jones polynomial for all classical knots. See [10, 21] for various virtual enhancements of the Jones polynomial. Other interesting virtual knot invariants can be found in [11, 21, 34] and elsewhere.

8.3 Virtual Unknotting

In Chapter 7, we delved into some unknotting operations for classical knots. In particular, we learned about the crossing change (CC) move that can be used to turn any knot diagram into a diagram of the unknot. Perhaps surprisingly, the CC move is *not* an unknotting operation for virtual knots. We can show this using one of our favorite examples.

Exercise 8.3.1. Consider the virtual knot diagram D in Figure 8.2.5. We computed $\mathbf{p}_t(K)$ for this virtual knot using D and saw that $\mathbf{p}_t(K) \neq 0$, thus proving that the virtual knot is nonclassical. In particular, we proved that the virtual knot is *not* the unknot. Show that, for every possible virtual knot diagram D' that is related to D by some number of CC moves, the intersection index polynomial is nonzero. Thus, D cannot be unknotted using CC moves.

Another famous example of a nonclassical virtual knot that cannot be unknotted with CC moves is **Kishino's knot**, pictured in Figure 8.3.1 and first introduced in [24]. Unlike the example in Figure 8.2.5, more powerful virtual knot invariants than J or $\mathbf{p}_t$ are needed to prove that Kishino's knot is nontrivial. (See, for instance, [4, 10].)

Figure 8.3.1: Kishino's knot.

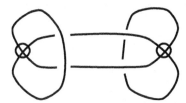

Exercise 8.3.2. Prove that neither the odd writhe nor the intersection index polynomial can distinguish Kishino's knot from the unknot.

If we take a closer look at Kishino's knot, we see that the reason this virtual knot diagram is nontrivially and virtually knotted is a consequence of the fact that certain Reidemeister-type moves are forbidden. In Exercise 8.1.7, you showed that four of the potential virtual Reidemeister moves fail to describe a virtual knot equivalence. Two of these moves are particularly interesting: moves (e) and (f). These two moves are commonly referred to as the **forbidden moves**. Notice that if either of the forbidden moves were allowable as virtual R-moves, Kishino's knot would be unknottable.

Exercise 8.3.3. Show that Kishino's knot can be unknotted using virtual Reidemeister moves together with one or both of the forbidden moves, i.e., moves (e) and (f) in Figure 8.1.7.

Since we are able to unknot Kishino's knot using forbidden moves, we might ask, are there *other* virtual knots that can be unknotted with forbidden moves? The answer is a resounding "Yes!" Sam Nelson proved the following theorem in [29].

Theorem 8.3.4. *If both of the forbidden moves are added to the collection of classical and virtual Reidemeister moves, then* every *virtual (and classical) knot can be unknotted.*

The way Nelson proved this fascinating result is actually quite simple. He showed that the Gauss diagrammatic forbidden moves along with Gauss diagrammatic Reidemeister moves can be used to rearrange all arrows in any Gauss diagram so that they become nonintersecting. As you proved in Exercise 4.2.7, such a Gauss diagram must represent an unknot. Let's reconstruct the details of Nelson's proof.

Exercise 8.3.5. Use the Gauss diagram schema you derived in Exercise 8.1.7 for forbidden moves (e) and (f) in Figure 8.1.7 along with Gauss diagrammatic R-moves from Figures 4.2.5, 4.2.4, and 8.3.2 to justify each of the moves in the Gauss diagram sequence shown in Figure 8.3.3. In other words, identify moves *i, ii, iii, iv,* and *v* if *n* is replaced by + and $-n$ is replaced by $-$. Similarly, identify moves *i, ii, iii, iv,* and *v* if *n* is replaced by $-$ and $-n$ is replaced by +.

Exercise 8.3.6. In Exercise 8.3.5, Gauss diagrammatic R-moves and forbidden moves were used to move the head of an arrow past the tail of an arrow in a Gauss diagram *when the arrows have the same sign.* Now, prove that the head of an arrow can be moved past the tail of an arrow that has the *opposite sign* (as shown in Figure 8.3.4) by providing a sequence of Gauss diagrammatic R-moves and forbidden moves similar to the one shown in Figure 8.3.3.

Before we put all of the pieces together and complete our proof, let's see how these derived forbidden moves are able to unknot several nontrivial virtual knots.

Figure 8.3.2: Four Gauss diagrammatic Reidemeister 3 moves that do not appear in Figure 4.2.5, but can be derived from the minimal generating set shown in the figure. (Did you find all four of these moves in Exercise 4.2.9?)

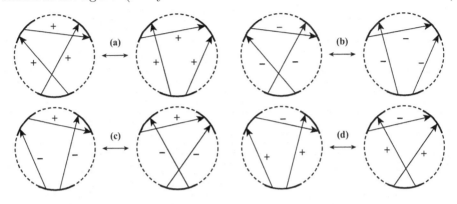

Figure 8.3.3: A sequence of Gauss diagrammatic Reidemeister and forbidden moves.

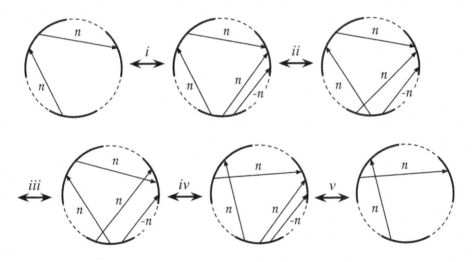

Exercise 8.3.7. Unknot the two nontrivial virtual knots in Figure 8.1.3 using the derived forbidden moves from Figures 8.3.3 and 8.3.4.

Exercise 8.3.8. Use Exercises 4.2.7, 8.3.5, and 8.3.6 to prove Theorem 8.3.4.

It is clear now why we don't allow forbidden moves to be used to

Figure 8.3.4: A move that can be derived from a sequence of Gauss diagrammatic Reidemeister and forbidden moves.

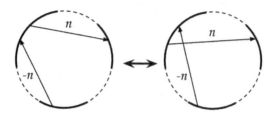

characterize virtual knot equivalence. Curiously, if just *one* of the two forbidden moves is added to the list of classical and virtual R-moves, the resulting knot theory isn't trivial. (Recall that we needed *both* forbidden moves in order to complete our proof above.) In fact, this new kind of knot theory is quite interesting! It is referred to as the theory of **welded knots** [18].

We just showed that the pair of forbidden moves can unknot any virtual knot. Just as in classical knot theory, there are many types of unknotting operations. Let's look at a different flavor of unknotting operation for virtual knots called *virtualization*. When a classical crossing in a virtual knot diagram is replaced by a virtual crossing, as in Figure 8.3.5, we call this a **virtualization** of the crossing.

Figure 8.3.5: A classical crossing becomes a virtual crossing via the virtualization operation.

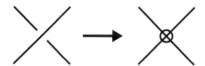

In Exercises 8.1.4 and 8.1.5, we proved (perhaps without realizing it) that virtualization is an unknotting operation. Indeed, if all or nearly all of the classical crossings in a virtual knot diagram are virtualized, then the diagram becomes the unknot. An interesting question emerges from this idea: can we unknot a given virtual knot diagram with fewer virtualization moves?

Exercise 8.3.9. Perform some number $n < 3$ of virtualization moves on the diagram of Kishino's knot in Figure 8.3.1 to transform the diagram into an unknot. Show, using virtual R-moves, that your virtualized diagram is indeed the unknot.

Given that the virtualization move is a virtual unknotting operation, we can define the virtual unknotting number of a virtual knot K.

Definition 8.3.10. The **virtual unknotting number**, $u_v(K)$, of K is the minimum number of virtualization moves that must be performed on a diagram D of K in order to produce the unknot. Here, the minimum is taken over all possible diagrams D of K.

Exercise 8.3.11. Make a conjecture about the value of $u_v(K)$ for the virtual knot K shown in Figure 8.1.5.

Exercise 8.3.12. Provide an infinite family of virtual knots with virtual unknotting number 1. (Hint: There are two parts to this problem. First, use the odd writhe or the intersection index polynomial to prove that all of the virtual knots in your family are distinct. Second, prove that each member of the family does indeed have virtual unknotting number 1.)

Finally, let's think about the virtual unknotting number more generally.

Exercise 8.3.13. In Section 7.2, we defined the trivializing number $t(D)$ of a knot diagram D via its Gauss diagram G_D. Since $t(D)$ is defined in terms of Gauss diagrams, its definition can be extended to virtual knots. Use the trivializing number to find an upper bound on the virtual unknotting number of a virtual knot.

Acknowledgments

We would like to thank Charles Livingston and Colin Adams for their thoughtful feedback on this manuscript. We are also grateful to Erin McNicholas, David Neel, and Leanne Robertson for providing helpful suggestions for improvement. Most especially, we would like to recognize our knot theory students—Elsa, Fintan, Hunter, Jacob, Justin, Leigh, Randi, Sean, Spencer, Taz, Ana, Kees, Megan, Rufei, Sarah, and Zach—who helped us to make significant improvements to the book throughout the writing and editing process as well as those students who inspired us to write this book in the first place.

Finally, we owe a debt of gratitude our wonderful families and friends. Thank you for being our cheerleaders! Your patience, support, and love are so greatly appreciated!

Index

Bibliography

[1] C. Adams, *The Knot Book: An Elementary Introduction to the Mathematical Theory of Knots*, American Mathematical Society, Providence, RI, 2004.

[2] H. Aida, *Unknotting operations of polygonal type*, Tokyo J. Math., **15**, (1992), 111–121.

[3] J. W. Alexander, *A lemma on systems of knotted curves*, Proc. Nat. Acad. Sci. USA, **9**, (1923), 93–95.

[4] A. Bartholomew and R. Fenn, *Quaternionic invariants of virtual knots and links*, J. Knot Theory Ramifications, **17**, (2008), 231–251.

[5] J. S. Birman and T. E. Brendle, *Braids: a survey*, arXiv:math/0409205v2, (2004), 1–17.

[6] S. A. Bleiler, *A note on unknotting number*, Math. Proc. Camb. Phil. Soc., **96**: 3, (1984), 469–471.

[7] Z. Cheng, *A transcendental function invariant of virtual knots*, arXiv:1511.08459, (2015), 1–12.

[8] S. Chumutov, S. Duzhin, J. Mostovoy, *Introduction to Vassiliev knot invariants*, Cambridge University Press, Cambridge, 2012.

[9] J. H. Conway, An enumeration of knots and links, and some of their algebraic properties, *Computational Problems in Abstract Algebra*, Pergamon Press, (1969), 329–358.

[10] H. A. Dye and L. H. Kauffman, *Virtual crossing number and the arrow polynomial*, J. Knot Theory Ramifications, **18**: 10, (2009), 1335–1357.

[11] R. Fenn, M. Jordan-Santana, and L. H. Kauffman, *Biquandles and virtual links*, Topology Appl., **145**, (2004), 157–175.

[12] J. Goldman and L. H. Kauffman, *Rational tangles*, Adv. in Appl. Math., **18**, (1997), 300–332.

[13] R. Hanaki, *Pseudo diagrams of knots, links, and spatial graphs*, Osaka J. Math., **47**, (2010), 863–883.

[14] A. Henrich, *A sequence of degree one Vassiliev invariants for virtual knots*, J. Knot Theory Ramifications, **19**: 4, (2010), 461–487.

[15] A. Henrich, N. MacNaughton, S. Narayan, O. Pechenik, and J. Townsend, *Classical and virtual pseudodiagram theory and new bounds on unknotting numbers and genus*, J. of Knot Theory Ramifications, **20**: 4, (2011), 625–650.

[16] V. F. R. Jones, *A polynomial invariant for knots via von Neumann algebras*, Bull. Amer. Math. Soc., **12**, (1985), 103–112.

[17] N. Kamada and S. Kamada, *Abstract link diagrams and virtual knots*, J. Knot Theory Ramifications, **9**: 1, (2000), 93–106.

[18] S. Kamada, *Braid presentations of virtual knots and welded knots*, Osaka J. Math., **44**: 2, (2007), 441–458.

[19] L. H. Kauffman, *A self-linking invariant of virtual knots*, arXiv:math/0405049, (2004), 1–36.

[20] L. H. Kauffman, *State models and the Jones polynomial*, Topology **26**: 3, (1987), 395–407.

[21] L. H. Kauffman, *Virtual knot theory*, Europ. J. Combinatorics, **20**, (1999), 663–691.

[22] L. H. Kauffman, S. Lambropoulou, and S. Jablan. *Introductory Lectures on Knot Theory: Selected Lectures Presented at the Advanced School and Conference on Knot Theory and Its Applications to Physics and Biology*, **46** World Scientific, (2011).

[23] L. H. Kauffman and S. Lambropoulou, *On the classification of rational tangles*, Adv. in Appl. Math., **33**: 2, (2004), 199–237.

[24] T. Kishino and S. Satoh, *A note on non-classical virtual knots*, J. Knot Theory Ramifications, **13**, (2004), 845–856.

[25] C. Livingston, *Knot theory*, Carus Mathematical Monographs **24**, Mathematical Association of America, Washington, DC (1993).

[26] H. Murakami, *Some metrics on classical knots*, Math. Ann., **270**, (1985), 35–45.

[27] Y. Nakanishi, *Replacements in the Conway third identity*, Tokyo J. Math., **14**, (1991), 197–203.

[28] Y. Nakanishi. *Unknotting numbers and knot diagrams with the minimum crossings*, Mathematics Seminar Notes, Kobe University, **11**, (1983), 257–258.

[29] S. Nelson, *Unknotting virtual knots with Gauss diagram forbidden moves*, J. Knot Theory Ramifications, **10**: 6, (2001), 931–935.

[30] M. Polyak, *Minimal generating sets of Reidemeister moves*, Quantum Topol., **1**, (2010), 399–411.

[31] K. Reidemeister, *Elementare begründung der knotentheorie.*, Abh. Math. Sem. Univ. Hamburg, **5**:1, (1927), 24–32.

[32] J. Roberts, *Knots Knotes*, unpublished lecture notes (2010), http://math.ucsd.edu/~justin/Roberts-Knotes-Jan2015.pdf

[33] D. Rolfsen, *Knots and Links*, Publish or Perish, Inc., Berkeley, Calif., 1976.

[34] J. Sawollek, *An orientation-sensitive Vassiliev invariant for virtual knots*, J. Knot Theory Ramifications, **12**: 6, (2003).

[35] M. G. Scharlemann, *Unknotting number one knots are prime*, Inventiones mathematicae, **82**: 1, (1985), 37–55.

[36] H. Schubert, *Knoten mit zwei Brüken*, Math. Zeitschrift, **65**, (1956), 133–170.

[37] A. Shimizu, *Region crossing change is an unknotting operation*, J. Math. Soc. Japan, **66**: 3, (2014), 693–708.

[38] P. G. Tait, *On knots I, II, III.* Scientific Papers, **1**, London: Cambridge University Press, (1900), 273–347.

[39] H. F. Trotter, *Non-invertible knots exist*, Topology, **2**: 4, (1963), 275–280.

[40] V. Siwach and P. Madeti, *Region unknotting number of 2-bridge knots,* J. Knot Theory Ramifications, **24**: 11, (2015), 1–20.